GALLIUM ARSENIDE DIGITAL CIRCUITS

THE KLUWER INTERNATIONAL SERIES
IN ENGINEERING AND COMPUTER SCIENCE
VLSI, COMPUTER ARCHITECTURE AND
DIGITAL SIGNAL PROCESSING

Consulting Editor
Jonathan Allen

Other books in the series:

Logic Minimization Algorithms for VLSI Synthesis. R.K. Brayton, G.D. Hachtel, C.T. McMullen, A. Sangiovanni-Vincentelli. ISBN 0–89838–164–9.
Adaptive Filters: Structures, Algorithms, and Applications. M.L. Honig, D.G. Messerschmitt. ISBN 0–89838–163–0.
Introduction to VLSI Silicon Devices: Physics, Technology and Characterization. B. El-Kareh, R.J. Bombard. ISBN 0–89838–210–6.
Latchup in CMOS Technology: The Problem and Its Cure. R.R. Troutman. ISBN 0–89838–215–7.
Digital CMOS Circuit Design. M. Annaratone. ISBN 0–89838–224–6.
The Bounding Approach to VLSI Circuit Simulation. C.A. Zukowski. ISBN 0–89838–176–2.
Multi-Level Simulation for VLSI Design. D.D. Hill, D.R. Coelho. ISBN 0–89838–184–3.
Relaxation Techniques for the Simulation of VLSI Circuits. J. White, A. Sangiovanni-Vincentelli. ISBN 0–89838–186–X.
A VLSI Architecture for Concurrent Data Structures. W.J. Dally. ISBN 0–89838–235–1.
Yield Simulation for Integrated Circuits. D.M.H. Walker. ISBN 0–89838–244–0.
VLSI Specification, Verification and Synthesis. G. Birtwistle, P.A. Subrahmanyam. ISBN 0–89838–246–7.
Serial Data Computation. S.G. Smith, P.B. Denyer. ISBN 0–89838–253–X.
Simulated Annealing for VLSI Design. D.F. Wong, H.W. Leong, C.L. Liu. ISBN 0–89838–256–4.
FET Modeling for Circuit Simulation. D. Divekar. ISBN 0–89838–264–5.
VLSI Placement and Global Routing Using Simulated Annealing. C. Sechen. ISBN 0–89838–281–5.
Adaptive Filters and Equalizers. B. Mulgrew, C.F.N. Cowan. ISBN 0–89838–285–8.
Computer-Aided Design and VLSI Device Development, Second Edition. K.M. Cham, S-Y. Oh, J.L. Moll, K. Lee, P. Vande Voorde, D. Chin. ISBN 0–89838–277–7.
Automatic Speech Recognition. K-F. Lee. ISBN 0–89838–296–3.
Speech Time-Frequency Representations. M.D. Riley. ISBN 0–89838–298–X.
A Systolic Array Optimizing Compiler. M.S. Lam. ISBN 0–89838–300–5.
Switch-Level Timing Simulation of MOS VLSI Circuits. V.B. Rao, D.V. Overhauser, T.N. Trick, I.N. Hajj. ISBN 0–89838–302–1.
VLSI for Artificial Intelligence. J.G. Delgado-Frias, W.R. Moore (Editors). ISBN 0–7923–9000–8.
Wafer Level Integrated Systems: Implementation Issues. S.K. Tewksbury. ISBN 0–7923–9006–7.
The Annealing Algorithm. R.H.J.M. Otten, L.P.P.P. van Ginneken. ISBN 0–7923–9022–9.
VHDL: Hardware Description and Design. R. Lipsett, C. Schaefer, C. Ussery. ISBN 0–7923–9030–X.
The VHDL Handbook. D. Coelho. ISBN 0–7923–9031–8.
Unified Methods for VLSI Simulation and Test Generation. K.T. Cheng, V.D. Agrawal. ISBN 0–7923–9025–3.
ASIC System Design with VHDL: A Paradigm. S.S. Leung, M.A. Shanblatt. ISBN 0–7923–9032–6.
BiCMOS Technology and Applications. A.R. Alvarez (Editor). ISBN 0–7923–9033–4.
Analog VLSI Implementation of Neural Systems. C. Mead, M. Ismail (Editors). ISBN 0–7923–9040–7.
The MIPS-X RISC Microprocessor. P. Chow. ISBN 0–7923–9045–8.
Nonlinear Digital Filters: Principles and Applications. I. Pitas, A.N. Venetsanopoulos. ISBN 0–7923–9049–0.
Algorithmic and Register-Transfer Level Synthesis: The System Architect's Workbench. D.E. Thomas, E.D. Lagnese, R.A. Walker, J.A. Nestor, J.V. Rajan, R.L. Blackburn. ISBN 0–7923–9053–9.
VLSI Design for Manufacturing: Yield Enhancement. S.W. Director, W. Maly, A.J. Strojwas. ISBN 0–7923–9053–7.
Testing and Reliable Design of CMOS Circuits. N.K. Jha, S. Kundu. ISBN 0–7923–9056–3.
Hierarchical Modeling for VLSI Circuit Testing. D. Bhattacharya, J.P. Hayes. ISBN 0–7923–9058–X.
Steady-State Methods for Simulating Analog and Microwave Circuits. K. Kundert, A. Sangiovanni-Vincentelli, J. White. ISBN 0–7923–9069–5.
Introduction to Analog VLSI Design Automation. M. Ismail, J. Franca. ISBN 0–7923–9071–7.
Mixed-Mode Simulation. R. Saleh, A.R. Newton. ISBN 0–7923–9107–1.
Automatic Programming Applied to VLSI CAD Software: A Case Study. D. Setliff, R.A. Rutenbar. ISBN 0–7923–9112–8.
Models for Large Integrated Circuits. P. Dewilde, Z.-Q. Ning. ISBN 0–7923–9115–2.
Gallium Arsenide Digital Circuits. O. Wing. ISBN 0–7923–9081–4.

GALLIUM ARSENIDE DIGITAL CIRCUITS

by

Omar Wing

Department of Electrical Engineering and
Center for Telecommunications Research
Columbia University

KLUWER ACADEMIC PUBLISHERS
Boston/Dordrecht/London

Distributors for North America:
Kluwer Academic Publishers
101 Philip Drive
Assinippi Park
Norwell, Massachusetts 02061 USA

Distributors for all other countries:
Kluwer Academic Publishers Group
Distribution Centre
Post Office Box 322
3300 AH Dordrecht, THE NETHERLANDS

Library of Congress Cataloging-in-Publication Data

Wing, Omar.
    Gallium arsenide digital circuits / by Omar Wing.
      p.   cm. — (The Kluwer international series in engineering and
computer science : 109. ISSN VLSI, computer architecture, and
digital signal processing.)
    Includes bibliographical references and index.
    ISBN 0-7923-9081-4 :
    1. Metal semiconductor field-effect transistors.   2. Modulation
-doped field-effect transistors.   3. Gallium arsenide
semiconductors.   4. Digital integrated circuits—Design and
construction—Data processing.   I. Title.   II. Series: Kluwer
international series in engineering and computer science ; SECS
109.   III. Series: Kluwer international series in engineering and
computer science.   VLSI. computer architecture, and digital signal
processing.
TK7871.95.W56   1990
621.381 '5284—dc20   90-4992
                                                            CIP

To

Camella, David, Gregory and Jeannette

Contents

Preface

Gallium Arsenide technology has come of age. GaAs integrated circuits are available today as gate arrays with an operating speed in excess of one Gigabits per second. Special purpose GaAs circuits are used in optical fiber digital communications systems for the purpose of regeneration, multiplexing and switching of the optical signals. As advances in fabrication and packaging techniques are made, the operating speed will further increase and the cost of production will reach a point where large scale application of GaAs circuits will be economical in these and other systems where speed is paramount.

This book is written for students and engineers who wish to enter into this new field of electronics for the first time and who wish to embark on a serious study of the subject of GaAs circuit design. No prior knowledge of GaAs technology is assumed though some previous experience with MOS circuit design will be helpful. A good part of the book is devoted to circuit analysis, to the extent that is possible for non-linear circuits. The circuit model of the GaAs transistor is derived from first principles and analytic formulas useful in predicting the approximate circuit performance are also derived. Computer simulation is used throughout the book to show the expected performance and to study the effects of parameter variations.

For circuit designers who look for design ideas, this book is a good source of reference of GaAs digital circuits of the various families, ranging from the simplest enhancement-depletion logic, to buffered logic, to source-coupled logic. For those who look for application ideas, we have included a chapter on subsystems design. Among the subsystems studied are the multiplexer and demultiplexer, cross-point switch, time-and-space switch, static random access memory, and all the major circuits of a repeater in an optical fiber communications system such as the photo-detector, clock extraction circuit, decision circuit and the laser driver. Selected references are given at the end on other useful circuits such as frequency dividers, parallel multipliers and A/D and D/A converters.

We begin in Chapter 1 with an introduction to the electronic properties of GaAs and describe its substantial speed advantage over Si. Chapter 2 is devoted to the derivation of a circuit model of the GaAs MESFET that is suitable for circuit simulation. A charge-based model is the point of departure, from which the current-voltage and the charge-voltage relations are derived.

Circuit analysis and design begins in Chapter 3 and we start with the enhancement-depletion (E/D) logic. We derive analytic expressions for the noise margins and delays and show what tradeoffs are possible. Chapter 4 deals with transmission gate logic. This important family, so extensively used in MOS circuits, is shown to have special problems in GaAs. Buffered logic of many different types are studied in Chapter 5 and we show how it overcomes some of the limitations of E/D logic. Source-coupled logic is separately treated in Chapter 7 because of its unique property that its performance is largely independent of threshold variation. Chapter 7 is the last chapter in which we present examples of subsystems design, as mentioned earlier.

The book would not have been possible without the support and encouragement of many individuals. I want to thank Bernard T. Murphy, formerly of AT&T Bell Laboratories, who gave me the opportunity to spend a sabbatical year at Bell Labs to work on GaAs integrated circuit design. I am grateful to S.S. Pei, N. Shah and A. Chandra of Bell Labs from whom I learned much about GaAs technology. Lastly I want to thank my students at Columbia who suffered through two drafts of the class notes on which this book is based. Their comments and suggestions helped clarify and improve the presentation. Many of the design examples came from their class projects. Finally, I wish to acknowledge the help of Mrs. Betty Lim who so skillfully constructed all of the figures and diagrams on her computer so that the book could be finished on time.

                                                        Omar Wing

GALLIUM ARSENIDE DIGITAL CIRCUITS

Chapter 1

Introduction

1.1 Gallium Arsenide

GaAs transistors have two distinct advantages over Si transistors: speed and power. For the same power dissipation, a GaAs circuit is usually faster, and at the same speed, the power in a GaAs circuit is usually lower. The speed advantage comes from the fact that the peak average electron velocity in intrinsic or doped GaAs is several times higher than in Si and it is reached at a much lower value of electric field, and hence with a lower supply voltage. Since the current density in a device is proportional to the electron velocity, the amount of current available to charge or discharge a capacitor in a GaAs device is much larger and the switching speed is therefore higher than in a Si device with the same dimension. In addition, a GaAs field effect transistor does not have any pn-junction around its drain and source terminals and therefore the interelectrode capacitance in a GaAs device is much smaller. Smaller capacitance and higher current density, combined with a smaller voltage swing in a GaAs transistor, contribute to the realization of low-power, high-speed circuits.

GaAs field effect transistor circuits exist that operate at several Gigabits per second (Gb/s), and as amplifiers, they have a unity gain frequency as high as tens of Giga-Hertz (GHz). In the early stages of development, GaAs devices were used as discrete components in microwave circuits, principally in satellite communication systems (from L-band or 1.5-1.6 GHz to K-band or 17-31 GHz). Recently, GaAs integrated circuits appear in digital fiber-optic communication systems as multiplexers, demultiplexers, and timing circuits [1-5].

At the same time, the demand for faster computers requires faster logic circuits. Studies are under way to determine the feasibility of using GaAs technology to build the next generation of computers. Experimental GaAs RAMs, parallel multipliers, microprocessors and other computer circuits with an access time or cycle time of a few nanoseconds or less have been reported [6-10].

Recently a new family of materials based on heterojunction engineering has emerged that promises even greater speed and higher frequency of operation. The most common among these is an AlGaAs/GaAs heterojunction in which a two-dimensional electron gas is formed at the interface of the two materials that have differenct band gaps. The electron mobility there is as much as three orders of magnitude larger than in Si. GaAs bipolar transistors are also possible in which the emitter-base junction is an AlGaAs/GaAs heterojunction. The discontinuity of the band gap at the junction gives rise to a large transistor beta and at the same time makes it possible to have a heavily doped base, thus reducing the base resistance and hence increasing the switching speed of the device.

In this book, we shall limit the study of GaAs digital circuits to GaAs MESFET (metal-semiconductor field effect transistor) circuits. The current-voltage characterisitcs of heterojunction field effect transistor (HFET) are similar to those of an MESFET and the design techniques of MESFET circuits apply directly to HFET circuits, so the latter will not be separately treated. Heterojunction bipolar transistor (HBT) circuits are still in development stage and will not be covered in this book.

In this chapter we will summarize the important properties of GaAs pertaining to device characteristics and circuit performance. Circuit models of the GaAs MESFET will be derived in the following chapter to show the important differences between a Si MOSFET and a GaAs MESFET. An empirical model suitable for circuit simulation purpose is given and it is used in all of the simulation exercises throughout the book. Logic circuits of various types are then presented, together with analysis techniques to gain a better insight into the principle of operation of the circuits. This is followed by examples of GaAs subsystems used in telecommunication applications.

1.2 Electronic Properties of GaAs

GaAs is a III-V compound with a direct band gap (the minimum of the conduction band - central valley - lined up with the maximum of the valence band). There is a second minimum (upper valley) at 0.36 eV above the central valley where the effective mass is five times greater than that at the central valley. The following is a brief summary of the electronic properties of GaAs compared with those of Si [11].

Property	GaAs	Si
Band gap	1.424 eV	1.12 eV
Effective mass		
Electron	0.063 m_0	0.33 m_0
Hole	0.090 m_0	0.16 m_0
Mobility		
Electron	6000 $cm^2/V-s$	1200 $cm^2/V-s$
Hole	350 $cm^2/V-s$	480 $cm^2/V-s$
Dielectric constant	13.1 ε_0	11.9 ε_0
Intrinsic resistivity	10^8 $\Omega-cm$	2.3×10^5 $\Omega-cm$
Minority carrier lifetime	10^{-8} sec	2.5×10^{-3} sec
Thermal conductivity	0.46 W/cm-C^o	1.5 W/cm-C^o
Intrinsic Debye length	2290 μm	24 μm
Intrinsic carrier concentration	$1.79 \times 10^6 cm^{-3}$	$1.45 \times 10^{10} cm^{-3}$
Effective density of states		
Conduction band	$4.7 \times 10^{17} cm^{-3}$	$2.8 \times 10^{19} cm^{-3}$
Valence band	$7 \times 10^{18} cm^{-3}$	$1.04 \times 10^{19} cm^{-3}$

Table 1.1   Electronic properties of GaAs and Si.

From the table, we see that GaAs has a relatively large band gap. This means that it can operate at relatively high temperatures. Its intrinsic carrier concentration is low so the material is a semi-insulator. Its resistivity is large so no special measures need be taken to provide isolation between devices on the chip. A direct band gap leads to short life time of minority carriers so that electron-hole pairs generated by radiation, for example, will recombine quickly before they can cause degradation of circuit performance. Its high electron mobility gives rise to high cut-off frequency of amplifiers and fast switching speed of digital circuits. On the other hand, its hole mobility is disproportionately low so that it is not practical to construct complementary circuits such as Si CMOS. In addition, GaAs does not have a native oxide so that it is not possible to build a MOS-like structure. Si_3N_4 is used as insulator on a GaAs substrate. It has a dielectric constant of 7.5 compared to 3.9 for SiO_2 so it has a higher capacitance for the same area. The thermal conductivity is relatively low so a wafer is often thinned to prevent excessive temperature rise. GaAs is also very brittle and yield loss through handling is significant.

Radiation hardness is a desirable property for space and military applications.

1.3 Velocity-Field Relation

In GaAs, as in most semiconductors, the electron velocity is a nonlinear function of the electric field. The steady state velocities in GaAs and Si have been measured and they have the form shown in Fig. 1.3.1.

The velocity in semi-insulating GaAs reaches its peak value of about 2.2×10^7 cm/s at approximately 3 kV/cm and decays to a saturation value of about 1.4×10^7 cm/s. The curves of Fig. 1.3.1 are somewhat misleading since the electrons in an actual device never reach the steady state. A fuller description of the velocity-field relation is obtained if we look at the transient behavior of velocity, i.e., velocity as a function of time and of transit distance in a device.

Before presenting the dynamic behavior, we note that electrons in a typical device do experience high field. If the gate length of a GaAs field effect transistor is $1\mu M$ and the drain to source voltage is 1.5 V. Then the electric field across the drain-source electrodes is 15 kV/cm.

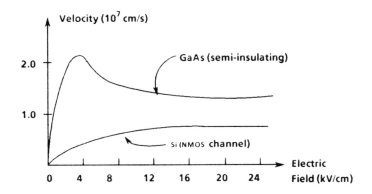

Fig. 1.3.1   Steady state electron velocity in GaAs and Si [12,13].

By numerical simulation of the motion of electrons in a field effect transistor, J.G. Ruch has obtained the transient behavior of electrons in Si and GaAs as summarized in Figs. 1.3.2 and 1.3.3 [14]. We see from Fig. 1.3.2 that the electrons in Si reach the steady state very quickly (< 1 ps) and that the steady state value increases monotonically towards a saturated value, as depicted in Fig. 1.3.1. The velocity in GaAs on the other hand attains a peak value several times the steady state, and steady state is not reached until several picoseconds have elapsed. Moreover, the steady state value of the velocity is not monotonic with respect to the electric field. At high fields, the steady state velocity actual decreases as the field strength increases.

Another way of looking at the results is to determine the distance over which the electrons must travel before the velocity reaches steady state. This is shown in Fig. 1.3.4 for Si and in Fig. 1.3.5 for GaAs.

Fig. 1.3.5 shows that the electrons in GaAs under high field never reach the steady state for a device with a gate length of 1 μM or less, and that the instantaneous drift velocity is actually higher than the steady state value. In contrast, the drift velocity in Si reaches its steady state

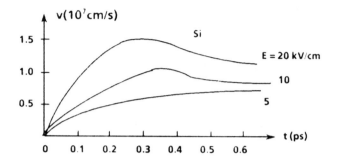

Fig. 1.3.2   Instantaneous drift velocity of electron in Si [14].

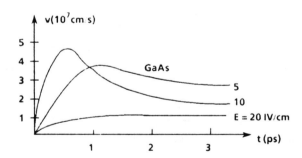

Fig. 1.3.3   Instantaneous drift velocity of electron in GaAs [14].

value after the electrons have traveled only about 0.1 μM, as seen from Fig. 1.3.4.

The reduction of the steady state drift velocity in GaAs as electric field increases can be explained as follows. Let E_C be some critical value of the electric field E. For $E < E_C$, all the conduction electrons are in the central valley where the effective mass is small and the electron

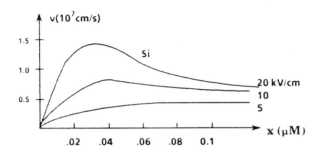

Fig. 1.3.4   Drift velocity over distance traveled by electron in Si [14].

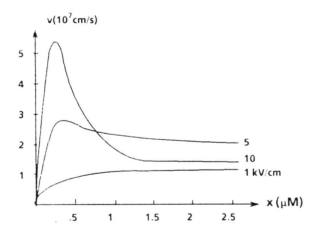

Fig. 1.3.5   Drift velocity over distance traveled by electron in GaAs [14].

velocity is high. For $E > E_C$, some of the electrons have sufficient energy to go into the upper valley where the effective mass is large and the velocity low. The average velocity at high field is therefore less.

1.4 GaAs Transistor Structures

The speed advantage of GaAs over Si must be somehow translated into fast and useful devices and circuits. Devices need be designed that will not only utilize the high velocity of the electrons to produce a high current density but will also provide a means to control the current so that such devices can act as switches or amplifiers. At the present time, three basic device structures have appeared. One is the MESFET. It is the simplest to fabricate. The second is HFET, also known as HEMT for high electron mobility transistor, or MODFET for modulation doped field effect transistor, or SDHT for selectively doped heterojunction transistor, or TEGFET for two-dimensional electron gas field effect transistor. The third is HBT. We shall describe the structure of each in the following.

MESFET

Fig. 1.4.1 shows a MESFET structure based on the recessed gate technology. An undoped buffer layer of GaAs is epitaxially grown on top a semi-insulating substrate. An active layer, with a donor concentration of about $10^{16} cm^{-3}$, is grown on top of the buffer layer. Lastly, an n+ layer with a concentration of about $10^{18} cm^{-3}$ is grown on top of the active layer. Etching is applied to remove all but the areas for the transistors. Ohmic contacts are then placed onto the n+ areas to form the drain and source electrodes. Then an area on each island is etched away for the placement of the gate electrode. By controlling the thickness of the active layer under the gate, an enhancement MESFET (normally off) or a depletion MESFET (normally on) can be made.

Note that in this structure, the distance between the source or drain electrode to the region under the gate is considerable so that the resistance along this path is significant. The consequence is that the circuit speed is reduced, as we will see later. The interelectrode distance can be reduced by a self-aligned process [15] and the structure is shown in Fig. 1.4.2. By ion-implantation, a layer of p-type GaAs is formed on the semi-insulating substrate. Then a layer of n-type GaAs is formed on top of the p-layer. The gate is then placed on the n-layer and ion implantation is used to create two n+ regions on each side of the gate. Ohmic contacts are then made on the n+ regions to form the drain and source terminals. Note that the n+ regions butt against the n region under the gate and the interelectrode resistance is much reduced.

HFET

Fig. 1.4.3 shows a recessed gate HFET structure in which an undoped GaAs layer, an undoped AlGaAs layer, an n+ AlGaAs and finally an n+ GaAs layer are grown in that order on a semi-insulating substrate. Islands of transistor areas are formed by etching and ohmic contacts are made on the n+ layer to form the drain and source electrodes. Lastly selective etching is applied to remove an area for the placement of the gate. Since the layers are usually made by molecular beam epitaxy, their thickness can be controlled very precisely (to within a few atomic layers), and devices made this way have more uniform characteristics than the recessed gate MESFET. The AlGaAs/GaAs interface is the heterojunction where the high mobility electrons reside.

There are many variants of the basic HFET structure. In some, the order of the layers is reversed to result in an "inverted HFET" [16]. In another, the AlGaAs layer is undoped but a thin and heavily doped layer is placed on the GaAs layer [17]. It is beyond the scope of this book to enumerate all of the new structures as they are being created almost at will at research laboratories.

HBT

Fig. 1.4.4 shows an HBT structure. It is similar to the Si bipolar transistor except that the emitter is made of AlGaAs so the base-emitter junction is a heterojunction of different types with different band gaps. The composition and doping concentration of the layers are shown in the figure.

It is not the purpose of this book to describe the details of the growth and processing techniques in the fabrication of GaAs integrated circuits. Interested readers are referred to reference [19] for further study.

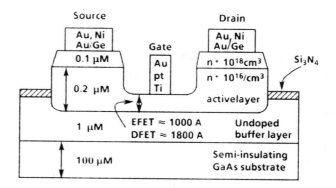

Fig. 1.4.1 A MESFET structure (not to scale).

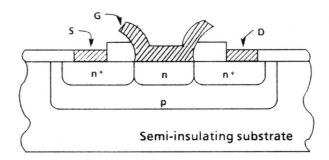

Fig. 1.4.2 A self-aligned MESFET structure [15].

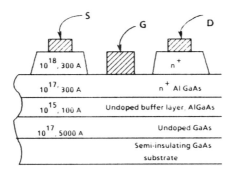

Fig. 1.4.3   HFET structure (not to scale).

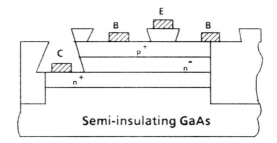

Layer	Al fraction	Thickness (A)	Doping (cm^{-3})
n+ cap	0	500	1×10^{19}
N emitter	0.25	2000	5×10^{17}
p+ base	0	600	2×10^{19}
n- collector	0	7000	5×10^{16}
n+ sub-collector	0	6000	5×10^{18}

Fig. 1.4.4 HBT structure and layer description [18].

1.5 Scope

The scope of this book, which is intended to be a small volume, is limited to the study of digital circuits for high speed (Gb/s) applications. Such related subjects as the design of GaAs monolithic microwave integrated circuits (MMIC) will not be covered. This is an important subject that deserves a separate, in-depth treatment [20]. Omitted, too, are opto-electronic devices and circuits (See [19].), and analog circuits such as operational amplifiers, comparators, and analog-to-digital converters [21, 22]. Advances in GaAs circuit design in all of these areas are regularly reported in references [23] and [24].

In this book, the emphasis will be on the analysis and design of digital circuits. Several families of logic circuits will be presented and studied in subsequent chapters, after the derivation of the circuit models of the GaAs MESFET, which is given next.

References

[1] P. Flahive, W. Clemetson, P. O'Connor, A. Dori, and S. C. Shunk, "A GaAs DCFL Chip Set for Multiplexer and Demultiplexer Applications at Gigabit/sec Data Rates," IEEE GaAs Integrated Circuit Symposium Technical Digest, pp. 7-10, October 1984.

[2] H. Nakamura, K. Tanaka, K. Inokuchi, T. Saito, Y. Kawakami, Y. Sano, M. Akiyama and K. Kaminishi, "2 GHz multiplexer and Demultiplexer Using DCFL/SBFL Circuit and the Precise Threshold Voltage Control Process," IEEE GaAs Integrated Circuit Symposium Technical Digest, pp. 151-4, October 1986.

[3] M. McDonald and G. McCormack, "A 12:1 Multiplexer and Demultiplexer Chip Set for Use in a Fiber Optic Communication System," IEEE GaAs Integrated Circuit Symposium Technical Digest, pp. 229-32, October 1986.

[4] T. Suzuki. S. Shikata, S. Nakajima, N. Hirakata, Y. Mikamura and T. Sugawa, "GaAs IC Family for High Speed Optical Communication Systems," IEEE GaAs Integrated Circuit Symposium Technical Digest, pp. 225-8, October 1986.

[5]   N. Kotera, K. Yamashita, Y. Hatta, T. Kinoshita, M. Miyakazi and M. Maeda, "Laser Driver and Receiver Amplifiers for 2.4 Gb/s Optical Transmission Using WSi-Gate GaAs MESFETs," IEEE GaAs Integrated Circuit Symposium Technical Digest, pp. 103-6, October 1987.

[6]   S. Notomi, Y. Asano, M. Kosugi, T. Nagata, K. Kosemura, M. Ono, N. Kobayashi, H. Ishiwari, K. Odani, T. Mimura and M. Abe, "A High Speed 1K x 4-Bit Static RAM Using 0.5 μm-gate HEMTs," IEEE GaAs Integrated Circuit Symposium Technical Digest, pp. 177-80, October 1987.

[7]   M. Ino, H. Suto, H. Kata and H. Yamazaki, "A 1.2 ns GaAs 4kb Read Only Memory Fabricated by 0.5 μm-gate BP-SAINT," IEEE GaAs Integrated Circuit Symposium Technical Digest, pp. 189-92, October 1987.

[8]   R. V. Gauthier, J. Weissman and B. E. Peterson, "A 150 MOPS GaAs 8-Bit Slice Processor," IEEE International Solid-State Circuits Conference Digest of Technical Papers, pp. 32-3, February 1988.

[9]   E. Delhaye, C. Rocher, M. Fichelson and I. Lecuru, "A 3.0 ns, 350 nW 8 x 8 Booth's Multiplier," IEEE GaAs Integrated Circuit Symposium Technical Digest, pp. 249-52, October 1987.

[10] V. M. Milutinovic and D. A. Fura, **Gallium Arsenide Computer Design**. Los Angeles, CA: IEEE Computer Society Press, 1988.

[11] S. Sze, **Physics of Semiconductor Devices**. New York, NY: Wiley, 1981.

[12] J. G. Ruch and G. S. Kino, "Measurements of Velocity-Field Characteristics of GaAs," Applied Physics Letters, Vol. 10, p.40, 1967.

[13] F. F. Fang and A. B. Fowler, "Hot Electron Effects and Saturation velocities in Silicon Inverse Layers," Journal of Applied Physics, pp. 1825-31, March 1970.

[14] J. G. Ruch, "Electron Dynamics in Short Channel FET," IEEE Transactions on Electron Devices, Vol. ED-19, pp. 652-4, May 1972.

[15] K. Yamasaki, N. Sato, and M. Hirayama,"Below 10ps Gate Operation with Buried-P-Layer SAINT FETs," Electronics Letters, Vol. 20, pp. 1029-31, December 1984.

[16] N. D. Cirillo, M. S. Shur, and J. K. Abrokwah, "Inverted GaAs/AlGaAs Modulation-Doped Field-Effect Transistors with Extremely High Transconductance," IEEE Electron Device Letters, Vol EDL-7, pp. 71-2, 1986.

[17] H. Hida, A. Okamoto, H. Toyoshima, and K. Ohata, "A High-Current Drivability i-AlGaAs/n-GaAs Doped Channel MIS-Like FET (DMT)," IEEE Electron Device Letters, Vol. EDL-7, pp. 625-6, November 1986.

[18] K. C. Wang, O. M. Asbeck, M. F. Chang, D. L. Miller, and G. J. Sullivan, "High-Speed MSI Current-Mode Logic Circuit Implemented with Heterojunction Bipolar Transistors," IEEE GaAs Integrated Circuit Symposium Technical Digest, pp. 159-62, October 1986.

[19] D. Ferry (Ed.), **GaAs Technology**. Indianapolis, IN: SAMS, 1985.

[20] R. Soares, J. Graffeuil and J. Obregon (Eds.), **Applications of GaAs MESFETs**. Dedham, MA: Artech House, 1983.

[21] K. de Graaf and K. Fawcett, "GaAs Technology for Analog-to-Digital Conversion (Invited)," IEEE GaAs Integrated Circuit Symposium Technical Digest, pp. 205-9, October 1986.

[22] J. Corcoran, K. Poulton, and T. Hornak, "A 1GHz 6b ADC System," IEEE International Solid State Circuits Conference Technical Digest, pp. 102-3, February 1987.

[23] IEEE GaAs Integrated Circuit Symposium Technical Digest. IEEE, 445 Hoes Lane, Piscataway, N.J. 08854.

[24] IEEE International Solid State Circuits Conference Technical Digest. IEEE 445 Hoes Lane, Piscataway, N.J. 08854.

Chapter 2

Circuit Models of the MESFET

2.1 Introduction

An idealized MESFET (metal-semiconductor field effect transistor) structure is shown in Fig. 2.1.1. An active layer of n-type GaAs is grown on top of a semi-insulating GaAs substrate. The drain and source electrodes make contact with the active layer through an n+ region under each. A third electrode, which is the gate, is placed directly on the active layer. From consideration of the energy band diagram at the metal-semiconductor interface, we find there is a depletion region under the gate whose height is controlled by a transverse electric field created by an applied gate voltage V_G. The undepleted region is called the channel. When a positive voltage V_D is applied to the drain with respect to the source, a longitudinal electric field is created that accelerates electrons in the channel from the source towards the drain. The resulting current in the channel, called the drain current, will depend on the gate voltage and the drain to source voltage. A MESFET is therefore a three-terminal device whose drain to source current is controlled by the voltage on the third electrode.

In this chapter, we will derive a quasi-static model of the MESFET. We first derive the static current-voltage relations from first principles. The charge-voltage relation for the charge on the gate electrode is derived next. With that, together with the continuity equation, we will derive a system of first order differential equations that describes, to first

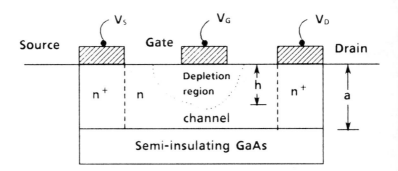

Fig. 2.1.1 An idealized MESFET structure.

order, the dynamic or transient behavior of the transistor. From this system, a capacitance matrix of the three-terminal device is obtained.

In addition to the drain current, there is current flowing from the gate into the semiconductor when the gate voltage is sufficiently positive with respect to the voltage along the channel. A distributed model of the gate current is given, which is then simplified to one that is represented by two-diodes.

It should be noted that the model given here relates the current-voltage-charge characteristics to the physical parameters of the device, so that from the model, one can predict how the device or circuit performance will change when the device parameters are varied. However, these relations are complicated implicit functions of the terminal voltages and numerical iterations are necessary to compute the currents and capacitance.

For the purpose of circuit simulation, a so-called empirical model is often used. In it, the current-voltage relation is approximated by an explicit function of the terminal voltages. This relation, however, is strictly mathematical and has little to do with the physics of the device and is therefore unsuitable for use in device characterization. Nevertheless, it will be used in this book as the model in the study of circuit performance.

2.2 Schottky Junction

We recall that the current flowing across a cross-sectional area A is proportional to the charge density and to the average velocity of the charge carriers crossing the area, namely,

$$I = Q \, A \, v \qquad (2.2.1)$$

where I is the current, Q the volume charge density, and v the velocity. In a MESFET, Q is the dopant concentration in the GaAs, A is the height of the channel times the width of the gate, and v is the electron velocity in GaAs, which is a function of the longitudinal electric field, as noted in Chapter 1.

The height of the channel is related to the height of the depletion region under the metal-semiconductor interface, known as a Schottky junction. In this section we will derive an expression relating the height of the depletion region to the gate voltage, and this expression will be used in the derivation of the drain current.

Consider a Schottky junction and its energy band diagram shown in Fig. 2.2.1. An n-type semiconductor is assumed. At thermal equilibrium and in the absence of an applied voltage across the junction, the Fermi levels of the metal and semiconductor must line up. Since there are many unfilled energy states in the metal, electrons in the semiconductor with sufficient energy will go over to the metal and leave an equal amount of positive charge in the semiconductor side. In addition, the junction is one of dissimilar materials and there will be a large number of "surface states" at the surface. Additional electrons from the semiconductor bulk will fill these surface states, further increasing the amount of positive charge in the immediate area near the junction. A depletion region is thus formed. The height h of the region is such that the potential built up by the positive charge is just high enough to prevent further movement of electrons into the metal or surface states.

For GaAs, the number of surface states at the junction is large so that the potential height (barrier height) on the metal side is approximately independent of the number of electrons absorbed in the surface states, and the barrier height V_B is said to be "pinned" to a constant value. From the band diagram, we see the barrier height is related to the built-in potential V_{bi} by:

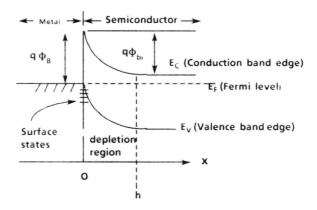

Fig. 2.2.1 Band diagram of metal/n-type semiconductor junction.

$$-qV_B = -qV_{bi} + E_C - E_F \qquad (2.2.2)$$

where E_C is the conduction band edge in the bulk semiconductor and E_F the Fermi level.

Depletion Height and Capacitance

Consider the case of zero gate bias first. In thermal equilibrium, the electron concentration n and hole concentration p are given by

$$n = N_C e^{-(E_C-E_F)/kT} \qquad (2.2.3)$$

$$p = N_V e^{-(E_F-E_V)/kT} \qquad (2.2.4)$$

where N_C and N_V are the effective densities of states for the conduction and valence bands, respectively, and $E_C(x)$ and $E_V(x)$ are the conduction and valence band edges, respectively, both functions of x. Inside the semiconductor, the variation of $E_C(x)$ and $E_V(x)$ are governed by Poisson's equation:

$$\frac{d^2E_C}{dx^2} = \frac{d^2E_V}{dx^2} = -q\frac{d^2V}{dx^2} = \frac{q\rho(x)}{\varepsilon} \qquad (2.2.5)$$

where $V(x)$ is the electrostatic potential, $\rho(x)$ the space charge density, and ε the permittivity of the semiconductor.

Assume all donors and acceptors are ionized. Then the space charge density is given by

$$\rho(x) = q\left[N_D(x) - N_A(x) - n(x) + p(x) \right] \qquad (2.2.6)$$

where $N_D(x)$ and $N_A(x)$ are the donor and acceptor densities, respectively.

For an n-type semiconductor, we assume $N_A=0$ and $p(x)=0$. Far away from the junction, the semiconductor is electrically neutral so that

$$\rho(\infty) = q(N_D - n(\infty)) = 0$$

Let

$$E_C(\infty) = E_{C\infty}$$

Then

$$n(\infty) = N_D = N_C e^{-(E_{C\infty}-E_F)/kT} \qquad (2.2.7)$$

where we have assumed that N_D is constant throughout (uniform doping).

The space charge density at any x can now be written as

$$\rho(x) = qN_D\left[1 - e^{-(E_C(x)-E_{C\infty})/kT} \right] \qquad (2.2.8)$$

Noting that

$$E_C(x) - E_{C\infty} = -qV(x) \qquad (2.2.9)$$

we find Poisson's equation to be

$$\frac{d^2V}{dx^2} = -\frac{qN_D}{\varepsilon}\left[1 - e^{qV/kT} \right] \qquad (2.2.10)$$

with boundary conditions:

$$V(\infty) = 0 \quad \text{and} \quad \frac{dV}{dx}\bigg|_{\infty} = 0 \qquad (2.2.11)$$

Eq. (2.2.10) cannot be solved analytically. For our purpose, we do not need the details of $V(x)$. The electric field at the surface will determine the depletion height and capacitance.

We note that

$$\frac{d^2V}{dx^2} = \frac{d}{dx}(\frac{dV}{dx}) = \frac{d}{dV}(\frac{dV}{dx})(\frac{dV}{dx})$$

Integrating Eq. (2.2.10) and imposing the boundary conditions that at $x=\infty$, $V=0$ and the electric field $E=-dV/dx=0$ there, we get

$$E(x) = -\left[\frac{2qN_D}{\varepsilon}(-V(x) - kT/q + kT/q \exp(qV(x)/kT)) \right]^{1/2} \quad (2.2.12)$$

Since from Eq. (2.2.9) $V(0) = -V_{bi}$, under zero gate bias, we get

$$E(0) = -\left[\frac{2qN_D}{\varepsilon}(V_{bi} - kT/q + kT/q \exp(-qV_{bi}/kT)) \right]^{1/2} \quad (2.2.13)$$

If a gate bias V_G is applied across the junction with the bulk semiconductor maintained at 0V, then we have

$$E(0) = -\left[\frac{2qN_D}{\varepsilon}(V_{bi}-V_G - kT/q + kT/q \exp(-q(V_{bi}-V_G)/kT)) \right]^{1/2}$$

$$(2.2.14)$$

In practice, $V_{bi}-V_G \gg kT/q$ and the exponential term can be neglected. The surface charge density Q_S is given by

$$Q_S = \varepsilon E(0)$$

and is negative to balance the positive charge in the depletion region. The "differential" gate capacitance is given by

$$C = \frac{dQ_S}{dV_G}$$

Taking derivative, we get

$$C = \left[\frac{\varepsilon q N_D}{2} \frac{1}{V_{bi}-V_G-kT/q} \right]^{1/2} \quad (2.2.15)$$

provided that $V_{bi}-V_G \gg kT/q$. This capacitance can be regarded as the capacitance of a parallel plate capacitor with a separation $h=\varepsilon/C$. So we have

$$h = \left[\frac{2\varepsilon}{qN_D}(V_{bi}-V_G-kT/q) \right]^{1/2} \quad (2.2.16)$$

It should be noted that if we neglect the kT/q term, the depletion height is the same as that derived with the assumption that the charge density is constant and equal to N_D for $0 \leq x < h$ and is zero elsewhere.

Eq. (2.2.16) shows that as we increase the gate bias positively, the depletion height shrinks and the channel height and therefore the current will increase. At the same time, Eq. (2.2.15) shows that the differential gate capacitance increases as the gate bias increases.

Barrier Heights

The Schottky barrier height for n-type GaAs is fairly independent of the type of metal that forms the junction, as the following table shows [1].

Semicond	Type	Ag	Al	Au	Pt	Ti	W
Si	n	0.78	0.72	0.80	0.90	0.50	0.67
Si	p	0.94	0.58	0.34		0.61	0.45
GaAs	n	0.88	0.80	0.90	0.84		0.80
GaAs	p	0.63		0.42			

Table 2.1 Barrier height in volts for various metals on different types of semiconductor.

It should be appreciated that for GaAs, the barrier height $V_B \gg kT/q$ in all cases.

Current Flow Across a Schottky Junction

When a forward bias is applied across a Schottky junction, current will flow due to the transport of electrons from the semiconductor into the metal. In addition tunneling of electrons through the barrier (for heavily doped semiconductors only) may occur. For GaAs with a moderately doped (10^{16}–$10^{17} cm^{-3}$) active layer, only the first kind of current is important and it is called the thermionic emission current. It will be denoted by I_G to stand for "gate current" and it is given by:

$$I_G = ART^2 \, e^{-qV_B/kT} \left[e^{qV_G/kT} - 1 \right] \qquad (2.2.17)$$

where A is the area of the junction, R the "Richardson's constant" given by

$$R = 4\pi q m' k^2 / h^3$$

where m' is the effective mass and h is Planck's constant. The derivation will be omitted and can be found in [1] for example.

From Eq. (2.2.17) we see that the gate current is similar to the p-n junction diode current. It becomes large when the bias V_G exceeds the barrier voltage V_B by several kT/q. The barrier voltage V_B will sometimes be referred to as the "turn-on" voltage of the junction.

A Schottky diode can be made by connecting the drain and source electrodes of a MESFET to a common node. Such a diode can be used as a "level shifter" as shown in Fig. 2.2.2.

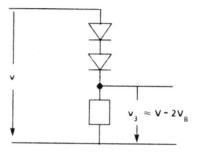

Fig. 2.2.2 Schottky diodes used as level shifters.

2.3 Drain Current

We now return to the MESFET structure, and we will derive an expression for the drain current as a function of the terminal voltages. The derivation is based on the work of [2,3,4]. To obtain an expression that is mathematically tractable, we must make approximations and assumptions that are reasonable from practical considerations. Even with these simplifications, the current expression is not explicit in the terminal voltages and iterations need be used to compute the current value for a given set of gate, drain, and source voltages.

Consider the MESFET structure shown in Fig. 2.3.1. The resistivity of the semi-insulating substrate is typically 10^8 Ω–cm. In the active layer, the dopant concentration is about 10^{16}–$10^{17} atoms/cm^3$, so its resistivity is about 0.01 Ω–cm. It follows that the substrate can be regarded as a perfect insulator so that the surface charge density on the n-GaAs/GaAs interface is zero. The dopant concentration of the n+ regions below the source and drain electrodes is about $10^{19} atoms/cm^3$ so its resistivity is about 0.0001 Ω–cm. We will assume that these regions are perfect conductors, and the n+/n junction built-in potential is assumed to be zero. Lastly, we assume the voltage drop from the source n+ region to the beginning of the channel, and from the drain n+ region to the end of the channel, is zero.

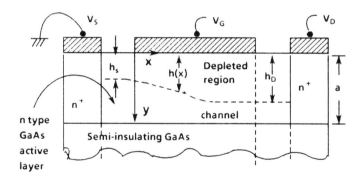

Fig. 2.3.1 MESFET structure showing the depletion region along the channel.

Let the gate voltage be V_G, the drain voltage be V_D and the source voltage be V_S. By convention, the terminal at a higher potential with respect to the other is called the drain so that $V_D \geq V_S$. Let the voltage along the channel be V. It is a function of both x and y. However, to simplify analysis, we assume that for the purpose of determining the depletion height under the gate, the voltage is a function of y only so that it satisfies a one-dimension Poisson's equation along y. This is reasonable if V varies slowly along x as it does near the source and far away from the drain, especially when the drain voltage is not too high.

Accordingly, from Eq. (2.2.16), the depletion height along the channel is

$$h = \left[\frac{2\varepsilon}{qN_D} (V_{bi} - V_G + V) \right]^{1/2} \tag{2.3.1}$$

in which we have neglected the kT/q term and used the fact that the voltage across the Schottky junction along x is $V_G - V$. The slow variation of V along x will be described as $V(x)$, and similarly $h(x)$.

Let the depletion height at the source end be h_S and that at the drain end be h_D. Then

$$h_S = \left[\frac{2\varepsilon}{qN_D} (V_{bi} - V_{GS}) \right]^{1/2} \tag{2.3.2}$$

$$h_D = \left[\frac{2\varepsilon}{qN_D} (V_{bi} - V_{GD}) \right]^{1/2} \tag{2.3.3}$$

where $V_{GS} \equiv V_G - V_S$ is the gate-to-source voltage and $V_{GD} \equiv V_G - V_D$ is the gate-to-drain voltage. Since $V_D > V_S$, $h_D > h_S$, as shown in Fig. 2.3.1.

Pinch-Off Voltage

From Eq. (2.3.3), we see that if the drain voltage with respect to the gate is sufficiently high (the Schottky junction is highly reversed biased near the drain), the depletion height h_D can reach the bottom of the active layer, i.e. $h_D = a$, and the region under the gate at this point is totally depleted. The channel is then said to be "pinched-off." Let $V_{GD(p)}$ be the gate-to-drain voltage at which pinch-off occurs. Then setting h_D to a and solving for $V_{GD(p)}$, we get

$$V_{GD(p)} = V_{bi} - \frac{qN_D}{2\varepsilon} a^2 \tag{2.3.4}$$

It is convenient to define the last term as the "pinch-off" voltage, i.e.

$$V_p \equiv \frac{qN_D}{2\varepsilon} a^2 \tag{2.3.5}$$

and we have

$$V_{GD(p)} = V_{bi} - V_p \tag{2.3.6}$$

Note that V_p is proportional to the second power of the thickness of the active layer and is therefore very sensitive to dimensional variations.

Threshold Voltage

As the drain voltage increases further, the depletion region extends towards the source so that part of the channel is pinched-off as shown in Fig. 2.3.2. It should be noted that for a fixed and sufficiently high drain voltage with respect to the source, pinch-off can occur by lowering the gate voltage. In fact, if the gate-to-source voltage is sufficiently low, the depletion height at the source end becomes a and the entire channel is totally depleted and devoid of any charge carrier. Under this condition, the drain current is zero (negligibly small) and we have cut-off. The value of V_{GS} at cut-off is called the **threshold voltage**, denoted by V_T, given by

$$V_T = V_{bi} - V_p = V_{bi} - \frac{qN_D}{2\varepsilon}a^2 \qquad (2.3.7)$$

The threshold voltage depends on device dimension, doping concentration, and the Schottky barrier height. By adjusting the thickness of the active layer a and doping concentration N_D, the threshold voltage can be made to have a positive or negative value, and the device becomes an enhancement or depletion type, respectively.

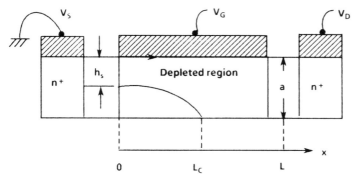

Fig. 2.3.2 Pinch-off of channel.

Velocity-Field Relation

The longitudinal electric field along the channel created by the drain-to-source voltage will accelerate the electrons towards the drain. Let the electron density be $n(x)$ and the average electron velocity be $v(x)$. Then from Eq. (2.2.1), the steady state drain current is given by

$$I_{DS} = qn(x)\,[a-h(x)]\,W\,v(x) \qquad (2.3.8)$$

where $[a-h(x)]$ is the height of the channel and W is its width. We shall assume the donor density is uniform and equal to N_D and that the donors are completely ionized so the charge carrier concentration $n(x)=N_D^+=N_D$.

The electron velocity $v(x)$ is a function of the longitudinal electric field E_x. For GaAs, the steady state velocity has a form shown in Fig. 1.1.3. Its peak value is about 2×10^7 cm/s and occurs at $E_x \approx 3KV/cm$. As we mentioned in the last chapter, the electrons never reach the steady state in a typical device. We note that for a $1\mu M$ gate and a drain-to-source voltage $V_{DS}=1.5V$, $E_x=15KV/cm$, so it is reasonable to expect that the electrons will attain their peak velocity somewhere in the channel before they reach the drain. Since the electrons do not stay in the channel long enough to attain their steady state velocity, which is smaller than the peak value, we argue that after they reach the peak velocity, they maintain the same velocity until they arrive at the drain. For our purpose, the velocity-field relation is therefore somewhat as shown in Fig. 2.3.3, and it will be approximated by [3]:

$$v(E) = \frac{\mu E}{1+\dfrac{E}{E_p}} \qquad 0 \le E \le E_C \qquad (2.3.9)$$

$$v(E) = \frac{\mu E_C}{1+\dfrac{E_C}{E_p}} \equiv v_s \qquad E > E_C \qquad (2.3.10)$$

where E_p is a parameter. For $E \ll E_p$, $v(E)=\mu E$ so that μ is the low field mobility. At a critical value $E=E_C$, v becomes constant at v_s.

From Eq. (2.3.8), it appears that if at some $x \le L$, where L is the length of the channel, $h(x)=a$, then the channel is pinched off there and $I_{DS}=0$. In fact, as the channel narrows, the electron velocity must increase to maintain current continuity throughout the channel. We have

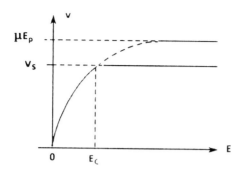

Fig. 2.3.3 Assumed velocity-field relation for the derivation of drain current.

assumed that as soon as the electrons attain a velocity v_s, they remain at this speed in the remainder of the channel. The only way to be consistent with current continuity is to argue that the channel is not completely closed but a small gap δ exists to let the electrons reach the drain. We can estimate the value of δ from Eq. (2.3.8) as follows.

$$I_{DS} = qN_D \, v_s \, \delta \, W$$

or

$$\delta = \frac{I_D}{qN_D \, v_s \, W} \qquad (2.3.11)$$

Assuming $N_D = 10^{17} cm^{-3}$, $v_s = 2 \times 10^7 cm/s$, and $I_D/W = 50 mA/mm$, we get $\delta \approx 160 A$, which is about 8% of the thickness of a typical active layer.

A better way to describe the conduction process in the channel is to note that as soon as the longitudinal electric field reaches a value E_C at some point $x = L_C$, the electron velocity for $x > L_C$ will be v_s, as depicted in Fig. 2.3.4. So the point $x = L_C$ marks the onset of velocity saturation and defines the boundary of two regions I and II. In I, the electron velocity is given by Eq. (2.3.9) and to the right, it is v_s. The longitudinal field at $x = L_C$ is E_C.

Two-Region Model

With reference to Fig. 2.3.1, in Region I, the longitudinal electric field is small compared to the transverse field. The depletion height $h(x)$

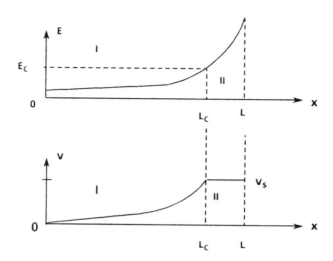

Fig. 2.3.4 Assumed field and velocity as functions of x.

is determined by a one-dimensional Poisson's equation. In Region II, the longitudinal field is no longer small and a two-dimension description of the potential is required.

Consider Region I first. With the electron velocity given by Eq. (2.3.9), the drain current is

$$I_{DS} = q N_D W \frac{\mu \dfrac{dV}{dx}}{1 + \dfrac{1}{E_p}\dfrac{dV}{dx}} \left[a - \left[\frac{2\varepsilon}{q N_D}(V_{bi} - V_G + V) \right]^{1/2} \right]$$

(2.3.12)

Let V_C be the channel voltage at $x=L_C$. Separating variables in Eq. (2.3.12) and integrating from $x=0$ to $x=L_C$ on one side and from $V=V_S$ to $V=V_C$ on the other, we get

$$I_{DS} = \frac{I_p}{1 + \dfrac{V_{CS}}{E_p L_C}} \left\{ \frac{V_{CS}}{V_p} + \frac{2}{3}\left[\left[\frac{V_{bi} - V_{GS}}{V_p} \right]^{3/2} - \left[\frac{V_{bi} - V_{GC}}{V_p} \right]^{3/2} \right] \right\}$$

(2.3.13)

where

$$I_p \equiv (qN_D)^2 \, a^3 \, \mu \, \frac{W}{L_C} \frac{1}{2\varepsilon} \tag{2.3.14}$$

$$V_{CS} \equiv V_C - V_S \tag{2.3.15a}$$

$$V_{GC} \equiv V_G - V_C \tag{2.3.15b}$$

In Eq. (2.3.13), L_C and V_C are unknown. We will derive expressions relating the two when we consider Region II.

Long-Channel, Low-Field Approximation

If the channel is long (gate length is large compared to the thickness of the active layer) and the drain voltage is low, the electric field along the channel never reaches the value E_C. In fact, if $E \ll E_p$ so that the velocity is given by $v = \mu E$, we obtain a long-channel, low-field approximation of the drain current:

$$I_{DS} = I_p \left\{ \left[\frac{V_{GS}}{V_p} + \frac{2}{3} \left[\frac{V_{bi} - V_{GS}}{V_p} \right]^{3/2} \right] - \left[\frac{V_{GD}}{V_p} + \frac{2}{3} \left[\frac{V_{bi} - V_{GD}}{V_p} \right]^{3/2} \right] \right\}$$

$$= f(V_{GS}) - f(V_{GD}) \tag{2.3.16}$$

where

$$f(x) \equiv I_p \left[\frac{x}{V_p} + \frac{2}{3} \left[\frac{V_{bi} - x}{V_p} \right]^{3/2} \right]$$

That is, I_{DS} is given by the difference of two identical functions, one with argument V_{GS} and one with V_{GD}. This is to be expected since the MESFET structure is symmetrical with respect to the source and drain.

Region II

We now turn to the derivation of I_{DS} in Region II. Here the electron velocity is constant at the saturation value v_s. At the boundary of the two regions, $V = V_C$ and the depletion height is

$$h_C = \left[\frac{2\varepsilon}{qN_D} (V_{bi} - V_G + V_C) \right]^{1/2} \tag{2.3.17}$$

which is assumed to be the same across this region. The drain current is

$$I_{DS} = qN_DWv_s(a - h_C) \qquad (2.3.18)$$

It must be continuous across the boundary. So setting I_{DS} as given by Eq. (2.3.13a) to that given by the above, we get one of the two necessary expressions relating the two unknowns V_C and L_C, i.e.

$$1 - \left[\frac{V_{bi} - V_{GC}}{V_p}\right]^{1/2} = \frac{V_p}{E_C L_C} \frac{1 + \dfrac{E_C}{E_p}}{1 + \dfrac{V_{CS}}{E_p L_C}}$$

$$\left\{\frac{V_{CS}}{V_p} + \frac{2}{3}\left[\left[\frac{V_{bi} - V_{GS}}{V_p}\right]^{3/2} - \left[\frac{V_{bi} - V_{GC}}{V_p}\right]^{3/2}\right]\right\} \qquad (2.3.19)$$

The second expression is obtained from solving for the potential in Region II and setting its value at the drain end to the applied drain voltage. In this region, the longitudinal electric field is large and and the potential $V(x,y)$ satisfies Poisson's equation:

$$\frac{\partial^2 V}{\partial x^2} + \frac{\partial^2 V}{\partial y^2} = -\frac{qN_D}{\varepsilon} \qquad (2.3.20)$$

and the boundary conditions

$$(1)\ V(x, 0) = V_G - V_{bi} \qquad (2.3.21a)$$

$$(2)\ \frac{\partial V}{\partial y}\bigg|_{y=a} = -E_y(x,a) = 0 \qquad (2.3.21b)$$

$$(3)\ \frac{\partial V}{\partial x}\bigg|_{x=L_C,\,y=a} = -E_x(L_C,a) = E_C \qquad (2.3.21c)$$

$$(4)\ V(L_C^+,y) = V(L_C^-,y) \qquad (2.3.21d)$$

The first condition says that the potential along the interface of the gate and the active layer in Region II is equal to the built-in potential reduced by the applied gate voltage. The second says the surface charge on the boundary of the semiconductor/semi-insulating interface is zero. Condition (3) requires the longitudinal field at $x=L_C$ be continuous. Condition (4) states that the potential across the two regions must be continuous.

To solve Poisson's equation, we decompose $V(x,y)$ into $V(x,y) = V_0(x,y) + V_1(x,y)$, in which $V_0(x,y)$ satisfies the homogeneous equation (Laplace equation) and $V_1(x,y)$ is a particular solution of Poisson's equation. That is

$$\frac{\partial^2 V_0}{\partial x^2} + \frac{\partial^2 V_0}{\partial y^2} = 0 \qquad (2.3.22)$$

with the boundary conditions:

(1) $V_0(x, 0) = 0$ (2.3.23a)

(2) $V_0(L_C, y) = 0$ (2.3.23b)

(3) $\left.\dfrac{\partial V_0}{\partial y}\right|_{y=a} = 0$ (2.3.23c)

(4) $\left.\dfrac{\partial V_0}{\partial x}\right|_{x=L_C, y=a} = E_C$ (2.3.23d)

The most general solution that satisfies Eq. (2.3.22) and the boundary conditions (1), (2) and (3) is

$$V_0(x,y) = \sum_{n=1}^{\infty} A_n \sin\frac{(2n-1)\pi}{2a} y \, \sinh\frac{(2n-1)\pi}{2a}(x - L_C) \qquad (2.3.24)$$

To satisfy condition (4), we must have

$$\sum_{n=1}^{\infty} A_n \frac{(2n-1)\pi}{2a} = E_C \qquad (2.3.25)$$

But this condition alone is not sufficient to determine the coefficients A_n. If the normal electric field along $x=L$ is known, then A_n can be determined from the Fourier series expansion of the field there. Since this field is not known, we must look for an approximation that is self-consistent. As explained in [2], A_n must diminish rapidly with n, for otherwise V_0 and E_x would grow exponentially along x via the sinh function. Accordingly, we assume that $A_n=0$ for all $n \neq 1$. So we get

$$V_0(x,y) = \frac{2a}{\pi} E_C \sin\frac{\pi}{2a} y \, \sinh\frac{\pi}{2a}(x - L_C) \qquad (2.3.26)$$

For the particular solution, we have

$$\frac{\partial^2 V_1}{\partial x^2} + \frac{\partial^2 V_1}{\partial y^2} = -\frac{\rho(y)}{\varepsilon} \qquad (2.3.27)$$

where $\rho(y) = N_D$ for $0 \leq y \leq h_C$ and $\rho(y) = 0$ for $h_C < y \leq a$. The boundary conditions on V_1 are:

(1) $V_1(x, 0) = V_G - V_{bi}$ \qquad (2.3.28a)

(2) $\left. \frac{\partial V_1}{\partial y} \right|_{y=a} = 0$ \qquad (2.3.28b)

(3) $\left. \frac{\partial V_1}{\partial x} \right|_{x=L_C} = 0$ \qquad (2.3.28c)

(4) $V_1(L_C^-, y) = V_1(L_C^+, y)$ \qquad (2.3.28d)

Conditions (1), (2) and (3) are obvious. Condition (3) says that we let $V_0(x,y)$ satisfy the boundary condition of Eq. (2.3.21c). Since $V_1(x,y)$ is any solution that satisfies the inhomogeneous equation and the boundary conditions, $V_1(x,y)$ need not be a function of x and the usual solution for a rectangular charge distribution $\rho(y)$ will suffice. So we have

$$V_1(x,y) = V_1(y) = V_C - \frac{qN_D}{2\varepsilon}(y - h_C)^2 \qquad (2.3.29)$$

The complete solution of the potential in Region II is then

$$V(x,y) = \frac{2a}{\pi} E_C \sin\frac{\pi}{2a}y \, \sinh\frac{\pi}{2a}(x - L_C) + V_C$$
$$- \frac{qN_D}{2\varepsilon}(y - h_C)^2 \qquad y \leq h_C \qquad (2.3.30)$$

$$V(x,y) = \frac{2a}{\pi} E_C \sin\frac{\pi}{2a}y \, \sinh\frac{\pi}{2a}(x - L_C) + V_C, \quad h_C < y \leq a \qquad (2.3.31)$$

Imposing the boundary condition at the drain, which is $V(L,a) = V_D$, we get, noting that $V_D - V_S = V_{DS}$ and $V_C - V_S = V_{CS}$:

$$V_{DS} = \frac{2a}{\pi} E_C \, \sinh\frac{\pi}{2a}(L - L_C) + V_{CS} \qquad (2.3.32)$$

This is the second expression we need that relates the unknowns L_C and

V_{CS}. Eqs. (2.3.32) and (2.3.19) will be solved simultaneously for L_C and V_{CS} or V_C. In fact, L_C can be eliminated and we get one nonlinear equation in one unknown, V_{CS}, given by:

$$\frac{V_p}{E_C L}(1 + \frac{E_C}{E_p}) \left\{ \frac{V_{CS}}{V_p} + \frac{2}{3} \left[\left[\frac{V_{bi} - V_{GS}}{V_p} \right]^{3/2} \right. \right.$$

$$\left. - \left[\frac{V_{bi} - V_{GS}}{V_p} + \frac{V_{CS}}{V_p} \right]^{3/2} \right] \right\}$$

$$- \left[1 - \left[\frac{V_{bi} - V_{GS}}{V_p} + \frac{V_{CS}}{V_p} \right]^{1/2} \right] \left\{ 1 + \frac{V_p}{E_p L} \frac{V_{CS}}{V_p} \right.$$

$$+ \frac{2a}{\pi L} \log \left[\frac{(V_{DS} - V_{CS})\pi}{2aE_C} + \left[\left[\frac{(V_{DS} - V_{CS})\pi}{2aE_C} \right]^2 + 1 \right]^{1/2} \right] \right\} = 0$$

(2.3.33)

which can be solved by Newton iteration. Once V_{CS} is determined, L_C is found from Eq. (2.3.32), or more explicitly:

$$L_C = L - \frac{2a}{\pi} \log \left[\frac{(V_{DS} - V_{CS})\pi}{2aE_C} + \left[\left[\frac{(V_{DS} - V_{CS})\pi}{2aE_C} \right]^2 + 1 \right]^{1/2} \right]$$

(2.3.34)

and the drain current is computed from Eq. (2.3.13). We will display the computed results shortly.

Saturation Drain voltage

At a given gate voltage V_{GS}, if we increase the drain voltage V_{DS} from zero, there will be a value of V_{DS} such that the electric field at the drain reaches the critical value E_C, and the electron velocity attains its saturation value v_s there. Let this value of drain voltage be denoted as V_{DSAT}. It is the value that marks the onset of velocity saturation. So for $V_{DS} < V_{DSAT}$, the electron velocity is less than v_s throughout the channel and there is but one region. For $V_{DS} = V_{DSAT}$, velocity saturation begins to take place at the drain and we have $L_C = L$ and $V_{CS} = V_{DSAT}$, and the second region just starts to form. For $V_{DS} > V_{DSAT}$, the channel is divided into two regions.

To find V_{DSAT}, we set $L_C = L$ and $V_{CS} = V_{DSAT}$ in Eq. (2.3.19), or alternately, we set $V_{CS} = V_{DS} = V_{DSAT}$ in Eq. (2.3.33), and solve for V_{DSAT} for each value of V_{GS}. The results are shown in Fig. 2.3.5 for a device with the following parameters: $E_C = 3kV/cm$, $v_s = 2 \times 10^7 cm/S$, $\mu = 8600 cm^2/V-S$, $N_D = 10^{17} cm-3$, $V_{bi} = 0.85V$, $V_T = -1V$, $L = 1\mu M$, and $W = 1000\mu M$.

It is seen that at $V_{GS} = V_T$, $V_{DSAT} = 0$ as it should, since the channel is completely cut-off. As V_{GS} increases from V_T, the channel widens and the drain voltage required to narrow the channel so that the electron velocity attains its saturation value, increases. At large positive values of V_{GS}, the depletion region disappears and the drain voltage to obtain velocity saturation is essentially independent of the gate voltage.

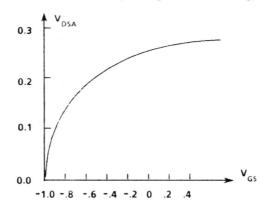

Fig. 2.3.5   Saturation drain voltage as a function of gate voltage.

Summary

We now summarize the drain current expressions for operation under the saturation and non-saturation conditions.

1. Non-saturation, $V_{GS} > V_T$ and $V_{DS} < V_{DSAT}$:

$$I_{DS} = \frac{I_p}{1 + \dfrac{V_{DS}}{E_p L}} \left\{ \frac{V_{DS}}{V_p} + \frac{2}{3} \left[\left[\frac{V_{bi} - V_{GS}}{V_p} \right]^{3/2} - \left[\frac{V_{bi} - V_{GS} + V_{DS}}{V_p} \right]^{3/2} \right] \right\}$$

$$(2.3.35)$$

where I_p is given by Eq. (2.3.14).

2. Saturation, $V_{GS} > V_T$ and $V_{DS} \geq V_{DSAT}$:

$$I_{DS} = \frac{I_p}{1 + \dfrac{V_{CS}}{E_p L_C}} \left\{ \frac{V_{CS}}{V_p} + \frac{2}{3} \left[\left[\frac{V_{bi} - V_{GS}}{V_p} \right]^{3/2} \right. \right.$$

$$\left. \left. - \left[\frac{V_{bi} - V_{GS} + V_{CS}}{V_p} \right]^{3/2} \right] \right\} \qquad (2.3.36)$$

where V_{CS} and L_C are determined from Eq. (2.3.33) and (2.3.34), respectively.

3. Cut-off, $V_{GS} \leq V_T$, $V_{DSAT} = 0$ and $I_{DS} = 0$.

Fig. 2.3.6 shows a set of computed current-voltage characteristics for a depletion type MESFET with the same set of parameters as before. It is seen that for a given V_{GS}, the current rises rapidly until it saturates at some value of V_{DS}. At the same time, for a given V_{DS}, the current increases supra-linearly as V_{GS} increases.

2.4 Gate Current

In the derivation of the drain current, we have assumed that the gate voltage is sufficiently low so that the thermionic emission current across the Schottky junction is negligible. In digital circuits, the gate voltage takes on extreme values, from a low value near the threshold voltage to a value comparable to the supply voltage. When the gate is driven to a large positive voltage, gate current will flow and as we will see later, it degrades the circuit performance in a significant way. This is particularly troublesome for enhancement transistors.

In this section, we shall derive an expression for the gate current. Consider the region under the gate in a transistor shown in Fig. 2.4.1. Let the terminal voltages be V_D, V_G, and V_S at the drain, gate and source, respectively. Let the corresponding terminal currents be I_D, I_G and I_S. It is convenient to write

$$I_D = I_{DS} - I_{GD} \qquad (2.4.1)$$

$$I_G = I_{GD} + I_{GS} \qquad (2.4.2)$$

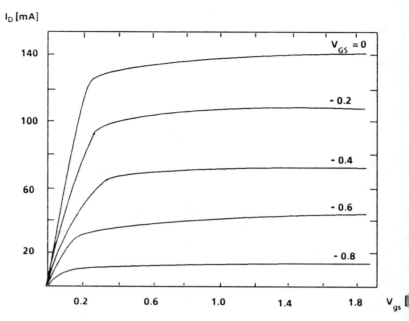

Fig. 2.3.6 Computed current-voltage characteristics of a
depletion MESFET.

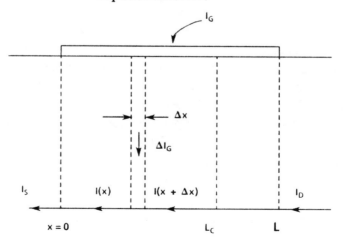

Fig. 2.4.1 MESFET structure pertaining to the derivation
of gate current.

$$I_S = I_{DS} + I_{GS} \tag{2.4.3}$$

where I_{DS} is the drain-to-source current in the absence of gate current, I_{GD} and I_{GS} are the part of the gate current that goes to the drain and source, respectively. The derivation given below follows the work of Chandra [5] who first derived a distributed model of the gate current for a heterojunction transistor.

Let $V(x)$ be the channel voltage at point x taken with respect to the source. Let $I(x)$ be the longitudinal channel current (in the -x direction). We assume the gate current consists of only thermionic emission current. Let the gate current density at x be $J_G(x)$. Then the gate current in the interval Δx is, from Eq. (2.2.17),

$$\Delta I_G(x) = J_o W \Delta x \left[e^{q(V_{GS} - V(x))/kT} - 1 \right] \tag{2.4.4}$$

where $J_o = ART^2 e^{-qV_B/kT}$, and W is the width of the gate. Over the interval $[x, x+\Delta x]$, we have

$$I(x) = I(x+\Delta x) + \Delta I_G(x)$$

from which we get

$$\frac{dI(x)}{dx} = -J_o W \left[e^{q(V_{GS} - V(x))/kT} - 1 \right] \tag{2.4.5}$$

This equation says that the longitudinal current decreases along x as we move from the source to the drain. The extra current as we go from the drain to the source comes from the distributed gate current along x. The longitudinal current $I(x)$ is the drift current which depends on the electron velocity. Using a 2-region model, we have for Region I:

$$I(x) = qN_D W \frac{\mu \dfrac{dV}{dx}}{1 + \dfrac{1}{E_p}\dfrac{dV}{dx}} \left\{ a - \left[\frac{2\varepsilon}{qN_D}(V_{bi} - V_{GS} + V(x)) \right]^{1/2} \right\} \tag{2.4.6}$$

or

$$\frac{dV(x)}{dx} = \frac{I(x)}{qN_D W \mu \left\{ a - \left[\dfrac{2\varepsilon}{qN_D}(V_{bi} - V_{GS} + V(x)) \right]^{1/2} \right\} - \dfrac{I(x)}{E_p}}$$

Eqs. (2.4.5) and (2.4.6) constitute a system of coupled differential equations in the unknowns $I(x)$ and $V(x)$. In principle, we can solve the equations once the boundary conditions are specified, which are $V(L_C) = V_C$ and $V(0) = V_S = 0$. As before, L_C is the point along the channel where the electron velocity equals the saturation velocity and the longitudinal electric field reaches E_C.

Let $I_C = I(L_C)$. We shall assume it is given by the total longitudinal current in Region II plus the total gate current in this region. That is

$$I_C = I_D + \int_{L_C}^{L} J_o W \left[e^{q(V_{GS} - V(x))/kT} - 1 \right] dx \qquad (2.4.7)$$

$$= I_D + I'_{GD}$$

It should be noted that if the drain voltage is so low that the field never reaches the value E_C, then there is but one region and $I_C = I_D$.

Writing Eq. (2.4.5) in integral form, we have

$$I(x) = I_C + \int_{x}^{L_C} J_o W \left[e^{q(V_{GS} - V(x'))/kT} - 1 \right] dx' \qquad (2.4.8)$$

Rewriting Eq. (2.4.6), we have

$$I(x)(dx + \frac{1}{E_p} dV) = q N_D W \mu \left\{ a - \left[\frac{2\varepsilon}{q N_D} (V_{bi} - V_{GS} + V) \right]^{1/2} \right\} dV$$

$$(2.4.9)$$

Substituting Eq. (2.4.8) into this equation and integrate from $x=0$ to $x = L_C$ on dx and from $V=0$ to $V = V_C$ on dV, we get

$$I_C(L_C + \frac{V_{CS}}{E_p}) + \int_{0}^{L_C} (1 + \frac{1}{E_p}\frac{dV}{dx}) \int_{x}^{L_C} J_o W \left[e^{q(V_{GS} - V(x'))/kT} - 1 \right] dx' dx$$

$$= q N_D W \mu \left\{ a V_{CS} - \frac{2}{3} \left[\frac{2\varepsilon}{q N_D} \right]^{1/2} \left[(V_{bi} - V_{GS} + V_{CS})^{3/2} - \right. \right.$$

$$\left. \left. (V_{bi} - V_{GS})^{3/2} \right] \right\} \qquad (2.4.10)$$

The right hand side is recognized to be $(L_C + \dfrac{V_{CS}}{E_p}) I_{DS}$, where I_{DS} is the drain-to-source current in the absence of gate current. We shall define the double integral as

$$(L_C + \frac{V_{CS}}{E_p})I_{GC} \equiv \int_0^{L_C}(1 + \frac{1}{E_p}\frac{dV}{dx})\int_x^{L_C}J_oW\left[e^{q(V_{GS}-V(x))/kT} - 1\right] dx'dx$$

(2.4.11)

Eq. (2.4.10) is then reduced to

$$I_{DS} = I_C + I_{GC}$$ (2.4.12)

This suggests an equivalent circuit given in Fig. 2.4.2 in which

$$I_S = I_{DS} + I_G - I_{GC} - I'_{GD}$$ (2.4.13)

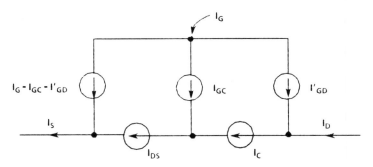

Fig. 2.4.2 Equivalent circuit of two-region model of
MESFET including gate current.

Integration of Eq. (2.4.11) by parts yields

$$I_{GC} = J_oW\int_0^{L_C}\left[\frac{x + \dfrac{V(x)}{E_p}}{L_C + \dfrac{V_{CS}}{E_p}}\right]\left[e^{q(V_{GS}-V(x))/kT} - 1\right] dx \quad (2.4.14)$$

If $V(x)$ is known for $0 \le x \le L$, I_{GC} can be evaluated (in principle at least), and I'_{GD} can be determined from the second term of Eq. (2.4.7) for Region II. Since I_{DS} is determined once L_C and V_{CS} are found, the source current I_S is determined from Eq. (2.4.13) with the total gate current found from

$$I_G = J_oW\int_0^{L_C}\left[e^{q(V_{GS}-V(x))/kT} - 1\right] dx + I'_{GD}$$ (2.4.15)

Following Chandra, we assume the channel voltage $V(x)$ to be a linear function of x:

$$V(x) = \frac{dV}{dx}\bigg|_{x=0} x = E_1 x \qquad 0 \le x \le L_C \qquad (2.4.16)$$

$$= \frac{V_{DS} - E_1 L_C}{L - L_C}(x - L_C) + E_1 L_C \qquad L_C \le x \le L \qquad (2.4.17)$$

Note that $V(x)$ satisfies the boundary conditions that $V(0) = 0$ and $V(L) = V_{DS}$. At $x = L_C$, it is likely that $E_1 L_C \ne V_{CS}$. But this is an internal point. As long as the external voltages and currents are self-consistent, any internal mismatch will be relatively unimportant.

Substituting Eq. (2.4.16) into Eq. (2.4.14), we get

$$I_{GC} = J_o W L_C \frac{1 + \dfrac{E_1}{E_p}}{1 + \dfrac{V_{CS}}{E_p L_C}}$$

$$\left\{ \frac{e^{qV_{GS}/kT}}{(\frac{qE_1 L_C}{kT})^2} \left[1 - (1 + \frac{qE_1 L_C}{kT}) e^{-qE_1 L_C/kT} \right] - \frac{1}{2} \right\} \qquad (2.4.18)$$

and the total gate current is given by

$$I_G = J_o W L_C \left\{ \frac{kT}{qE_1 L_C} e^{qV_{GS}/kT} \left[1 - e^{-qE_1 L_C/kT} \right] - 1 \right\} + I'_{GD} \qquad (2.4.19)$$

where

$$I'_{GD} = J_o W (L - L_C) \left\{ \frac{kT}{q(V_{DS} - E_1 L_C)} e^{q(V_{GS} - E_1 L_C)/kT} \right.$$

$$\left. \left[1 - e^{-q(V_{DS} - E_1 L_C)/kT} \right] - 1 \right\} \qquad (2.4.20)$$

It should be noted that I_{GC}, I_G and I'_{GD} are all functions of E_1, and we can write Eq. (2.4.13) as

$$I_S = I_{DS} + I_G(E_1) - I_{GC}(E_1) - I'_{GD}(E_1) \qquad (2.4.21)$$

where I_{DS} has been derived previously:

$$I_{DS} = \frac{qN_D W \mu}{L_C + \dfrac{V_{CS}}{E_p}} \left\{ aV_{CS} - \frac{2}{3} \left[\frac{2\varepsilon}{qN_D} \right]^{1/2} \left[(V_{bi} - V_{GS} + V_{CS})^{3/2} \right. \right.$$

$$\left. \left. - (V_{bi} - V_{GS})^{3/2} \right] \right\} \qquad (2.4.22)$$

Next, to be consistent, E_1 must satisfy the requirement that at $x = 0, I(0) = I_S$ and $dV/dx = E_1$, namely from Eq. (2.4.6):

$$E_1 = \frac{I_S}{qN_D W \mu \left[a - \left[\dfrac{2\varepsilon}{qN_D}(V_{bi} - V_{GS}) \right]^{1/2} \right] - \dfrac{I_S}{E_p}} \qquad (2.4.23)$$

Eqs. (2.4.23) and (2.4.21) can be solved by iteration. To this end, I_S can be simplified to be

$$I_S^0 = I_{DS} + J_o WL_C \frac{kT}{qE_1 L_C} e^{qV_{GS}/kT} \qquad (2.4.24)$$

which is obtained by dropping all terms multiplied by $e^{-qE_1 L_C/kT}$ and neglecting all terms that are smaller than $e^{qV_{GS}/kT}$. Substituting Eq. (2.4.24) into Eq. (2.4.23), we get a quadratic equation in E_1, which is solved to obtain the first iterate of E_1. This is then substituted into Eq. (2.4.21) to get the first iterate of I_S. Eq. (2.4.23) is then used to get the next iterate of E_1 and so on until the iteration converges.

Once E_1 and I_S are found, I_G, I_{GC} and I'_{GD} are determined. We now have a complete description of the gate currents at the terminals as prescribed by Eqs. (2.4.1-3) in which

$$I_{GD} = I_{GC} + I'_{GD} \qquad (2.4.25)$$

and

$$I_{GS} = I_G - I_{GC} - I'_{GD} \qquad (2.4.26)$$

Fig. 2.4.3 shows the computed results for a device with the same set of parameters as given in Sec.2.3.

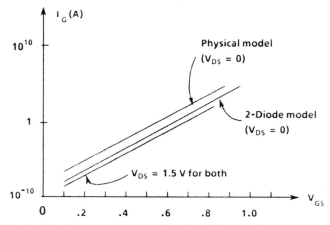

Fig. 2.4.3   Gate current as a function of gate voltage
for different values of drain voltages.

It is seen that the gate current is approximately an exponential function of the gate voltage. This suggests a simplification of the gate current model. In fact, the gate currents at the source and drain can be approximated by the currents of two diodes as shown in Fig. 2.4.4. With the diode parameters properly adjusted, the total gate current is seen to fit that computed from the physical model very well, as evident from Fig. 2.4.3.

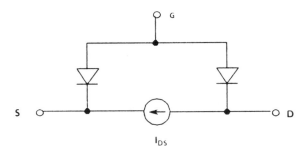

Fig. 2.4.4   Two-diode model of MESFET.

2.5 Capacitive Currents

So far we have derived the expressions for the static or DC currents at the terminals of a MESFET in terms of the terminal voltages. If the voltages vary slowly with time, we assume the terminal currents will follow the changes of the voltages, and at each instant of time, the values of the currents can be computed from the values of the voltages at that time, using the DC formulas for the current. However, when the rate of change of the voltages is large, we expect there will be capacitive currents at the terminals owing to the presence of surface charge on the gate and to the charge in the depletion region along the channel. In this section, we shall derive expressions for these capacitive currents.

From the point of view of circuit analysis and simulation, it is most convenient to postulate a terminal description as follows:

$$\begin{bmatrix} i_D(t) \\ i_G(t) \\ i_S(x) \end{bmatrix} = \begin{bmatrix} I_{DS} \\ 0 \\ -I_{DS} \end{bmatrix} + \begin{bmatrix} -I_{GD} \\ I_G \\ -I_{GS} \end{bmatrix} + \frac{d}{dt} \begin{bmatrix} Q_D \\ Q_G \\ Q_S \end{bmatrix} \qquad (2.5.1)$$

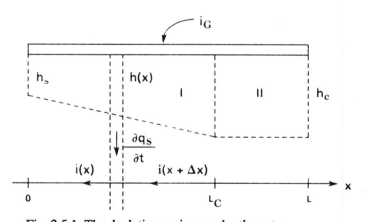

Fig. 2.5.1 The depletion regions under the gate.

with the directions of the currents shown in Fig. 2.5.1. In this equation, I_{DS} is the DC drain to source current in the absence of gate current. I_{GD} and I_{GS} are the part of the gate current that goes to the drain and source, respectively, and I_G is the total gate current. Q_D, Q_G, and Q_S are the "terminal" charges whose rate of change accounts for the capacitive currents at the terminals. Since $i_D + i_G + i_S = 0$, we must have $I_G = I_{GD} + I_{GS}$ as derived in the last section and

$$Q_D + Q_G + Q_S = 0 \qquad (2.5.2)$$

namely, the terminal charge is conserved at all times.

For convenience, we rewrite Eq. (2.5.1) as

$$\begin{bmatrix} i_D \\ i_G \\ i_S \end{bmatrix} = \begin{bmatrix} I_D \\ I_G \\ -I_S \end{bmatrix} + \frac{d}{dt} \begin{bmatrix} Q_D \\ Q_G \\ Q_S \end{bmatrix} \qquad (2.5.3)$$

where I_D, I_G, and I_S are the DC or slowly varying part of the respective terminal currents.

Consider Fig. 2.5.1 which shows the two depletion regions under the gate. As the terminal voltages change with time, the depletion height $h(x)$ and hence the amount of charge on the gate will change. The rate of change of the charge gives rise to a displacement current at the gate, which adds to the total channel current and emerges at the drain and source terminals. Let $q_s(x,t)$ be the surface charge density on the gate. Then the current continuity equation becomes

$$\frac{\partial i}{\partial x} = -W \frac{\partial q_s}{\partial t} \qquad (2.5.4)$$

In integral form, we have, noting $i(0,t) = -i_S(t)$,

$$i_S(t) = -i(x,t) - W \int_0^x \frac{\partial q_s}{\partial t} dx' \qquad (2.5.5)$$

Integrating over the channel and using integration by parts on the double integral, we get

$$i_S(t) = -\frac{1}{L} \int_0^L i(x,t)dx - \frac{d}{dt} \left[W \int_0^L (1 - \frac{x}{L}) q_s(x,t) dx \right] \qquad (2.5.6)$$

The equation is in the form of Eq. (2.5.3) and we define the quantity inside the brackets as the terminal charge at the source:

$$Q_S = -W \int_0^L (1 - \frac{x}{L}) q_s(x,t) dx \qquad (2.5.7)$$

In a similar manner, we find the terminal charge at the drain to be

$$Q_D = -W \int_0^L \frac{x}{L} q_s(x,t) dx \qquad (2.5.8)$$

The total charge on the gate is

$$Q_G = W \int_0^L q_s(x,t) dx \qquad (2.5.9)$$

and it is equal to $-Q_D - Q_S$ as required.

In a two-region model, the surface charge density at any point x is given by, for Region I:

$$q_s = -qN_D h(x) \qquad (2.5.10)$$
$$= -\sqrt{2\varepsilon q N_D (V_{bi} - V_{GS} + V(x))}$$

where we have used Eq. (2.3.1) for $h(x)$, and for Region II:

$$q_s = -\varepsilon E_C \sinh[\frac{\pi}{2a}(x - L_C)] - qN_D h_c \qquad (2.5.11)$$

which is obtained from Eq. (2.3.30) by evaluating $\partial V / \partial y$ at $y = 0$.

In order to proceed, we need to know $V(x)$ in Region I. Presumably, we could use the same solution technique as in the derivation of the gate current. But the expressions become intractable. We argue that since the terminal charge is the integral of the surface charge, which is proportional to the depletion height $h(x)$, it will be fairly insensitive to the fine structure of $h(x)$. Accordingly, we assume $h(x)$ to be a linear function of x in Region I, and we get

$$q_s = -qN_D \left[\frac{h_C - h_S}{L_C} x + h_S \right] \qquad (2.5.12)$$

with

$$h_C = \left[\frac{2\varepsilon}{qN_D} (V_{bi} - V_{GS} + V_{CS}) \right]^{1/2}$$

and

$$h_S = \left[\frac{2\varepsilon}{qN_D} (V_{bi} - V_{GS}) \right]^{1/2}$$

Substituting into Eq. (2.5.8) for $0 \leq x \leq L_C$ and using Eq. (2.5.11) for Region II, we get

$$Q_D = WqN_DL_C\left[\frac{1}{3}h_C\frac{L_C}{L} + \frac{1}{6}h_S\frac{L_C}{L}\right] \tag{2.5.13}$$

$$+ WqN_Dh_C(L - L_C)(\frac{L + L_C}{2L})$$

$$+ W\varepsilon E_C\frac{2a}{\pi}\left[\cosh\frac{\pi}{2a}(L - L_C) - \frac{2a}{\pi L}\sinh\frac{\pi}{2a}(L - L_C) - \frac{L_C}{L}\right]$$

$$Q_G = -WqN_DL_C\left[\frac{h_C}{2} + \frac{h_S}{2}\right] - WqN_Dh_C(L - L_C) \tag{2.5.14}$$

$$- W\varepsilon E_C\frac{2a}{\pi}\left[\cosh\frac{\pi}{2a}(L - L_C) - 1\right]$$

and

$$Q_S = WqN_DL_C\left[\frac{h_C}{2}(1 - \frac{2}{3}\frac{L_C}{L}) + \frac{h_S}{2}(1 - \frac{1}{3}\frac{L_C}{L})\right] \tag{2.5.15}$$

$$+ WqN_Dh_C(L - L_C)\left[1 - \frac{L + L_C}{2L}\right]$$

$$+ W\varepsilon E_C\frac{2a}{\pi}\left[\frac{2a}{\pi L}\sinh\frac{\pi}{2a}(L - L_C) - \frac{L - L_C}{L}\right]$$

It should be appreciated that h_C, h_S, and L_C are functions of the terminal voltages, so when the voltages are specified, the terminal charges Q_D, Q_G and Q_S can be computed.

Secondly, if there is but one region, that is, the device is not under velocity saturation, then $L_C=L$, $V_C=V_D$ and $h_C=h_D$. The charge expressions are simplified to be:

$$Q_D = WqN_DL\left[\frac{1}{3}h_D + \frac{1}{6}h_S\right] \tag{2.5.16}$$

$$Q_G = WqN_DL\left[\frac{1}{2}h_D + \frac{1}{2}h_S\right] \tag{2.5.17}$$

$$Q_S = W q N_D L \left[\frac{1}{6} h_D + \frac{1}{3} h_S \right] \tag{2.5.18}$$

2.6 Charge-Based Model for Circuit Simulation

Eq. (2.5.1) gives a quasi-static description of the terminal behavior of a MESFET. The slowly varying part is accounted for by the DC current and the fast changing part by the capacitive currents. This description is most suitable for circuit simulation.

In any circuit simulation program, the transient response of a circuit is computed by solving numerically at discrete time points, t_k, $k=1,2,...$, a system of nonlinear differential equations which are in fact the node equations of the circuit. The solution is usually based on an implicit integration method such as the Backward Euler method or the trapezoidal method. In the former case, each derivative term is replaced by a backward difference term, namely

$$\frac{dx}{dt} \approx \frac{x(t_{k+1}) - x(t_k)}{h}$$

where h is the step size. For the case of a MESFET, the set of terminal equations describing the transient behavior becomes at $t = t_{k+1}$:

$$\begin{bmatrix} i_D(t_{k+1}) \\ i_G(t_{k+1}) \\ i_S(t_{k+1}) \end{bmatrix} = \begin{bmatrix} I_{DS}(t_{k+1}) \\ 0 \\ -I_{DS}(t_{k+1}) \end{bmatrix} + \begin{bmatrix} -I_{GD}(t_{k+1}) \\ I_G(t_{k+1}) \\ -I_{GS}(t_{k+1}) \end{bmatrix} + \frac{1}{h} \begin{bmatrix} Q_D(t_{k+1}) - Q_D(t_k) \\ Q_G(t_{k+1}) - Q_G(t_k) \\ Q_S(t_{k+1}) - Q_S(t_k) \end{bmatrix} \tag{2.6.1}$$

Each term in the equation is a current term and the equation suggests that a MESFET can be replaced by a "companion circuit" for the purpose of circuit simulation, valid for $t = t_{k+1}$, as shown in Fig. 2.6.1. Each current term, including those from the terminal charges, is a function of the terminal voltages so that we can write, for example,

$$I_{DS}(t_{k+1}) = I_{DS}(V_D(t_{k+1}), V_G(t_{k+1}), V_S(t_{k+1})) \tag{2.6.2}$$

$$Q_D(t_{k+1}) = Q_D(V_D(t_{k+1}), V_G(t_{k+1}), V_S(t_{k+1})) \tag{2.6.3}$$

and etc. When these expressions are substituted into Eq. (2.6.1), each terminal current of the device is now a function of the node voltages at t_{k+1},

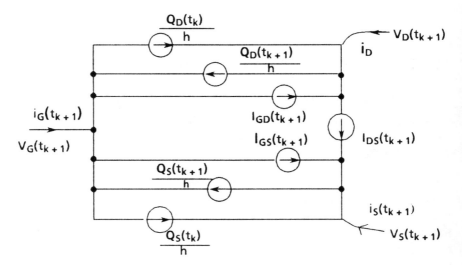

Fig. 2.6.1 Companion circuit model of a MESFET.

which are the variables of the node equations of the circuit. When the node equations are assembled for the circuit, we have a system of non-linear algebraic equations in the unknowns of the node voltages. The equations are solved by Newton iteration to obtain the node voltages at t_{k+1}. We then proceed to the next time point and repeat the solution process after each current and charge term has been updated by the node voltages just computed, until the solution has been computed for the desired number of time points.

It should be emphasized that at any time point, charge is conserved in each device. Secondly, though the currents associated with the terminal charges are capacitive, no capacitance is computed. (Actually, the Jacobian of the system of algebraic equations need be computed and those terms associated with the terminal charges have a unit of capacitance. However, terms of the Jacobian are usually computed by taking differences.)

2.7 Capacitance-Based Model

While the charge-based model is most suitable for the purpose of numerical circuit simulation, it is not useful in deriving circuit properties about the MESFET. To circuit designers, it is most convenient to represent the high frequency or high speed transient behavior in terms of capacitance, and if possible, they prefer to deal with an equivalent circuit composed of resistors, capacitors and simple controlled sources.

We begin with Eq. (2.5.1) and note that the system of equations is not independent because $i_D + i_G + i_S = 0$. We take the first and last equations and refer all voltages with respect to the gate. The terminal charges will be expressed in terms of the gate-to-drain and gate-to-source voltages, V_{GD} and V_{GS}, respectively. We get from Eq. (2.5.1),

$$\begin{bmatrix} i_D \\ i_S \end{bmatrix} = \begin{bmatrix} I_{DS} \\ -I_{DS} \end{bmatrix} + \begin{bmatrix} -I_{GD} \\ -I_{GS} \end{bmatrix} + \begin{bmatrix} \dfrac{\partial Q_D}{\partial V_{GD}} & \dfrac{\partial Q_D}{\partial V_{GS}} \\ \dfrac{\partial Q_S}{\partial V_{GD}} & \dfrac{\partial Q_S}{\partial V_{GS}} \end{bmatrix} \begin{bmatrix} \dfrac{dV_{GD}}{dt} \\ \dfrac{dV_{GS}}{dt} \end{bmatrix} \qquad (2.7.1)$$

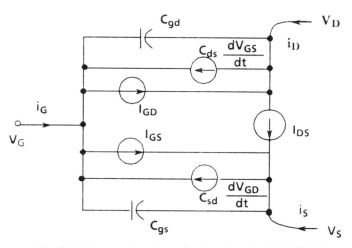

Fig. 2.7.1 Large signal equivalent circuit of MESFET.

This equation suggests an equivalent circuit shown in Fig. 2.7.1 where

$$C_{gd} \equiv -\frac{\partial Q_D}{\partial V_{GD}} \qquad C_{ds} \equiv \frac{\partial Q_D}{\partial V_{GS}} \qquad (2.7.2)$$

$$C_{sd} \equiv \frac{\partial Q_S}{\partial V_{GD}} \qquad C_{gs} \equiv -\frac{\partial Q_S}{\partial V_{GS}} \qquad (2.7.3)$$

all being functions of the voltages. The two controlled sources are unusual in that the current is capacitive, being proportional to the derivative of a voltage across another terminal pair. The proportionality constants C_{ds} and C_{sd} have a unit of capacitance and are called the transcapacitance. Note that they are not equal to each other in general and the circuit is therefore nonreciprocal with respect to the drain and source terminals.

2.8 Parasitic Elements

The circuit models derived so far are based on certain simplifying assumptions. A more realistic model must include parasitic elements as shown in Fig. 2.8.1, in which R_D and R_S account for the finite resistance of the drain and source electrodes and the voltage drop from the electrodes to the channel. R_G accounts for the resistance of the gate electrode and C_{GD}, C_{GS} and C_{DS} are interelectrode capacitances external to the device. The box labeled "intrinsic model" is the circuit of Fig. 2.6.1 if circuit simulation is being done, or it is the circuit of Fig. 2.7.1 if circuit analysis is under consideration.

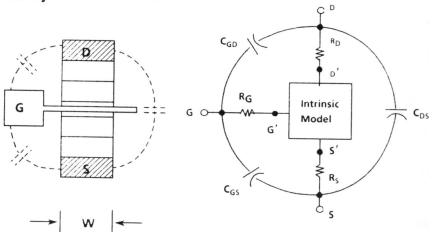

Fig. 2.8.1 Circuit model including parasitic resistances
and interelectrode capacitance.

The series resistances R_D and R_S can be reduced by scaling as the width of the device, W, is increased since W accounts for the cross-sectional area of the electrode over which the drain or source current flows. However, when W is increased, the drain current is also increased by the same factor (to first order) so that the voltage drop across R_D or R_S remains about the same. The drop across R_S is detrimental to circuit performance since it effectively reduces the net gate-to-source voltage and less drain current is available from the device for the same external applied gate-to-source voltage.

To effectively reduce R_D and R_S, the distance between the electrode and the channel must be shortened and the ohmic contact between the n^+ electrode and the metal terminal must be improved. A self-aligned fabrication process will help realize the former.

The gate electrode is usually connected to the external circuit via a metal contact at one end. In a typical device, $W \gg L$ and the gate current flows along the electrode of length W and cross-sectional area proportional to L. R_G is therefore proportional to W and inversely proportional to L. Since R_G is in series with the gate "capacitance", its effect is important during the time when the gate voltage changes rapidly. The effective gate voltage will not follow the applied gate voltage instantaneously but must rise or fall exponentially according to an "RC" time constant proportional to W^2.

As the width of the device increases, the interelectrode capacitances C_{GD}, C_{GS} and C_{DS} will increase. These quantities can be determined by solving a three-dimensional Laplace equation for a system of conductors embedded in a multilayer dielectric medium [6].

2.9 Small Signal Model

A circuit model for high frequency, small signal applications can be derived from that of Fig. 2.7.1. In practice, the gate is biased at a point where the gate current is negligible and the drain is biased at a point where the drain current is at saturation. When I_{DS} is expressed as a function of V_{GS} and V_{DS} and is expanded in Taylor series about an operating point V_{GS}^o and V_{DS}^o, we get

$$I_{DS} \approx I_{DS}(V_{DS}^o, V_{GS}^o) + \frac{\partial I_{DS}}{\partial V_{DS}}\bigg|_o (V_{DS} - V_{DS}^o) + \frac{\partial I_{DS}}{\partial V_{GS}}\bigg|_o (V_{GS} - V_{GS}^o)$$

$$(2.9.1)$$

where the partial derivatives are evaluated at the operating point and are recognized to be the drain conductance g_D (sometimes called the output conductance) and transconductance g_m, respectively:

$$g_D \equiv \frac{\partial I_{DS}}{\partial V_{DS}}\bigg|_o , \quad g_m \equiv \frac{\partial I_{DS}}{\partial V_{GS}}\bigg|_o \qquad (2.9.2)$$

To first order, both g_D and g_m are proportional to the width W of the device and are both functions of the terminal voltages. The operating point is usually chosen where g_m is largest and at the same time g_D is smallest (i.e. at drain current saturation).

The equivalent circuit becomes as shown in Fig. 2.9.1 in which the internal terminal voltages are V_D', V_G', and V_S'. The capacitances C_{gd}, C_{gs}, C_{sd}, and C_{ds} are evaluated at the operating point.

2.10 Empirical Model

From the previous sections, it is evident that the MESFET model derived so far is much too complicated for use in circuit design. In fact, it is too complicated even for circuit simulation. Each time the terminal voltages are changed, the drain current, gate current, and terminal charges must be re-evaluated and the evaluation requires that we solve a number of nonlinear algebraic equations by iteration. As the transient solution of a circuit is computed by numerical method, time point by time point, these currents and charges must be computed for each transistor over a number of terminal voltage values at each time point. This clearly requires a great deal of computer time and circuit simulation becomes very costly.

On the other hand, the model, which will be called the analytic model, is based on physical principles and it shows how the current-voltage characteristics of the transistor depend on the device parameters. The model provides a means to compute the sensitivity of circuit performance to changes in device dimensions, doping concentration, barrier

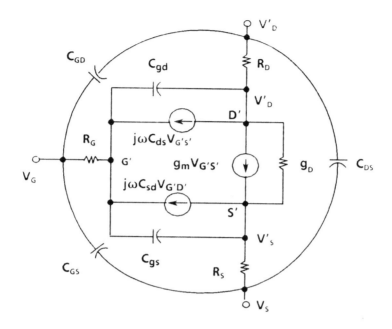

Fig. 2.9.1 Small signal, sinusoidal steady state equivalent
circuit of MESFET.

height, or layer thickness. Such a model is needed to understand the properties of the device and to determine the theoretical limits of its capabil-
ities.

Nevertheless, it is highly desirable to have a description of the
current- voltage characteristics in an explicit functional form. The model
should capture the essence of the device characteristics and be simple
enough to make it possible to derive analytic expressions of the circuit
performance for use in analysis and optimization of circuit design.

Current-Voltage Characteristics

Let us first examine the current-voltage characteristics of a MES-
FET and see how one might find a functional description of them. Figure 2.10.1 shows a typical set of current-voltage characteristics of a
depletion type GaAs MESFET. Those of an enhancement type are
shown in Figure 2.10.2. For a given gate-to-source voltage V_{GS}, the drain

current I_{DS} is a monotonically increasing function of the drain voltage V_{DS} that exhibits saturation at large values of V_{DS}. This suggests a tanh x type function.

Fig. 2.10.3 shows the drain current as a function of gate voltage for a set of fixed drain voltages. The current increases approximately quadratically as V_{GS} increases. The rate of increases is reduced as V_{GS} increases further. This suggests a function that is quadratic at low values of its argument and is less than quadratic for larger values.

When the gate voltage exceeds the Schottky barrier voltage, the gate becomes conducting and part of the gate current goes to the drain to reduce the net drain current, to the extent that for low values of V_{DS}, the drain current becomes negative, as is evident in Figure 2.10.2. The gate current can be accounted for by two diodes connecting the gate to the source and drain, respectively. The circuit model is shown in Fig. 2.10.4 in which we have added a series resistance to each of the device terminals to account for the resistance of the gate, of the source-to-channel and drain-to-channel voltage drops.

As evident from Fig. 2.10.3, the device also has a cut-off characteristic, namely, when V_{GS} is less than the threshold voltage V_T, the drain current is essentially zero, and the device can be regarded as a voltage-controlled switch. It is the cut-off and saturation characteristics that are utilized in the design of digital circuits. The two extremes signify minimum and maximum conduction of a transistor, as controlled by its gate voltage when the latter assumes a logically low or high value.

Many simplified models have been proposed. All are based on fitting the measured current characteristics to a functional description of the current. One of the most popular models is that of Statz [7] and it has been installed in SPICE.

In the Statz model, the drain current is given by

$$I_{DS} = \beta \frac{(V_{GS}-V_T)^2}{1 + b(V_{GS}-V_T)}(1 + \lambda V_{DS})\tanh \alpha V_{DS} \qquad (2.10.1)$$

where β is, according to [8]:

$$\beta = \frac{2\varepsilon\mu v_s W}{a(\mu V_p + 3v_s L)} \qquad (2.10.2)$$

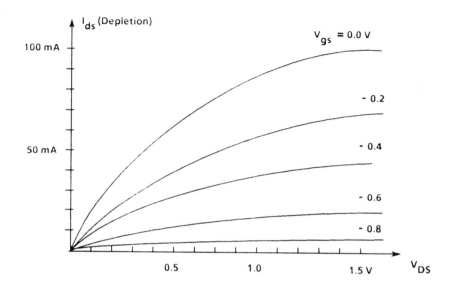

Fig. 2.10.1 Current-voltage characteristics of a depletion MESFET.

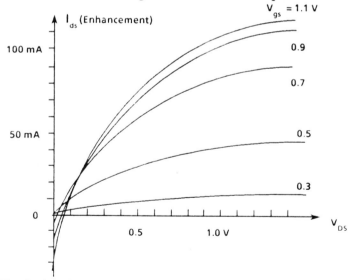

Fig. 2.10.2 Current-voltage characteristics of an enhancement MESFET.

in which ε is the permittivity, μ the low field mobility, v_s the electron saturation velocity, W the gate width, a the thickness of the active layer, L the gate length, and

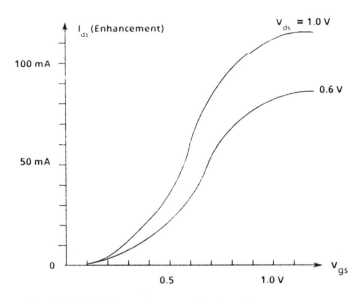

Fig. 2.10.3 Drain current as a function of gate-to-source
voltage of a MESFET, showing the cut-off characteristics.

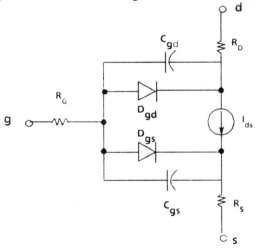

Fig. 2.10.4 Simplified equivalent circuit of a MESFET.

$$V_p = \frac{qN_d a^2}{2\varepsilon} \tag{2.10.3}$$

is the pinch-off voltage. N_d is the doping concentration of the active
layer.

In Eq. (2.10.1), the second term accounts for the dependence on the gate voltage. The expression in the denominator, with b as a parameter, describes the compression of the drain current when the gate voltage is large. The third term accounts for the finite slope of the drain current when the drain voltage is large and the slope is controlled by the parameter λ. The last term describes the drain current in both the linear and saturation regions with α as a parameter. Sometimes, the tanh x term is approximated by a polynomial and the drain current is given by

$$I_{DS} = I_{DSAT}(1 + \lambda V_{DS})[1 - (1 - \alpha V_{DS}/3)^3] \qquad (2.10.4)$$

for $V_{DS} \leq 3/\alpha$ and

$$I_{DS} = I_{DSAT}(1 + \lambda V_{DS}) \qquad (2.10.5)$$

for $V_{DS} > 3/\alpha$. I_{DSAT} is the "saturation current" given by

$$I_{DSAT} = \beta \frac{(V_{GS} - V_T)^2}{1 + b(V_{GS} - V_T)} \qquad (2.10.6)$$

Thus α determines the onset of the "knee" of the current characteristics. As shown in [7], the parameters of the Statz model can be adjusted to match the measured data almost exactly.

Charge and Capacitance

To complete the circuit model, we need to account for the capacitances at the terminals. For circuit simulation, as we saw from Sec. 2.6, it is the change of the charge as we go from time point t_k to t_{k+1} that describes the displacement current. It is not necessary to compute the capacitance. We argue that in a MESFET, there is but one physical charge and that is the total charge on the gate. The displacement currents at the drain and source will have to be derived from this charge alone.

The gate charge Q_G is a function of the gate-to-source and gate-to-drain voltages. As before, we assume that the gate-to-source and gate-to-drain displacement currents can be included in the terminal description as

$$i_D = I_{DS} - I_{GD} - \frac{dQ_{GD}}{dt} \qquad i_G = 0 + I_G + \frac{dQ_G}{dt} \qquad (2.10.7)$$

$$i_{GS} = -I_{DS} - I_{GS} - \frac{dQ_{GS}}{dt}$$

where the displacement currents are computed from

$$\frac{dQ_{GD}}{dt} \approx \frac{\Delta Q_{GD}}{h} \qquad (2.10.8)$$

$$\frac{dQ_{GS}}{dt} \approx \frac{\Delta Q_{GS}}{h}$$

$$\frac{dQ_{G}}{dt} \approx \frac{\Delta Q_{G}}{h}$$

and the incremental changes will be approximated as

$$\Delta Q_{GD} = \frac{1}{2}\Big[Q_G(V_{DS}(t_{k+1}), V_{GS}(t_{k+1})) - Q_G(V_{DS}(t_k), V_{GS}(t_{k+1})) \qquad (2.10.9)$$

$$+ Q_G(V_{DS}(t_{k+1}), V_{GS}(t_k)) - Q_G(V_{DS}(t_k), V_{GS}(t_k)) \Big]$$

$$\Delta Q_{GS} = \frac{1}{2}\Big[Q_G(V_{DS}(t_{k+1}), V_{GS}(t_{k+1})) - Q_G(V_{DS}(t_{k+1}), V_{GS}(t_k)) \qquad (2.10.10)$$

$$+ Q_G(V_{DS}(t_k), V_{GS}(t_{k+1})) - Q_G(V_{DS}(t_k), V_{GS}(t_k)) \Big]$$

$$\Delta Q_G = Q_G(V_{DS}(t_{k+1}), V_{GS}(t_{k+1})) - Q_G(V_{DS}(t_k), V_{GS}(t_k)) \qquad (2.10.11)$$

$$= \Delta Q_{GD} + \Delta Q_{GS}$$

It is seen that the sum of the displacement currents is zero.

It remains to find an expression for Q_G. In keeping with the spirit of making the model simple and explicit, the gate charge is approximated by the following function:

$$Q_G = 2C_{gs0}V_{bi}\left[1 + \left(1 - \frac{V_{eff1}}{V_{bi}} \right)^{1/2} \right] + C_{gd0}V_{eff2} \qquad (2.10.12)$$

$$V_{eff1} = \frac{1}{2}\left\{ V_{GS} + V_{GD} + \sqrt{(V_{GS} - V_{GD})^2 + \Delta^2} \right\} \qquad (2.10.13)$$

$$V_{eff2} = \frac{1}{2}\left\{ V_{GS} + V_{GD} - \sqrt{(V_{GS} - V_{GD})^2 + \Delta^2} \right\} \qquad (2.10.14)$$

where C_{gs0} and C_{gd0} are the gate-to-source and gate-to-drain capacitances, respectively, under zero gate bias, i.e. $V_{GS} = 0$. The capacitances under a general bias condition will be taken to be

$$C_{GD} = \frac{\partial Q_G}{\partial V_{GD}} \qquad (2.10.15)$$

$$C_{GS} = \frac{\partial Q_G}{\partial V_{GS}} \qquad (2.10.16)$$

The parameter Δ is chosen to smooth the transition of the capacitances as one becomes the other at $V_{DS} = 0$ when the drain becomes the source and vice versa. A value equal to $1/\alpha$ seems to work well.

In Eq. (2.10.12) , when $V_{eff\,1}$ exceeds V_{bi}, the expression breaks down, and it is suggested that it be replaced by

$$Q_G = C_{gs0} \left\{ 2V_{bi} \left[1 - \left[1 - \frac{V_{max}}{V_{bi}} \right]^{1/2} \right] + \frac{V_{eff\,1} - V_{max}}{\left[1 - \frac{V_{max}}{V_{bi}} \right]^{1/2}} \right\}$$

$$+ C_{gd0} V_{eff\,2} \qquad (2.10.17)$$

for $V_{eff\,1} \geq V_{max}$, with V_{max} a parameter. A suitable value is 0.5V.

Lastly, when the gate-to-source voltage goes beyond cutoff, i.e., $V_{GS} < V_T$, as often happens in circuit simulation, V_{max} is replaced by V_{new} where

$$V_{new} = \frac{1}{2} \left[V_{eff\,1} + V_T + \sqrt{(V_{eff\,1} - V_T)^2 + \delta^2} \right] \qquad (2.10.18)$$

Again, the parameter δ is used to smooth the transition of the capacitance C_{GS} from a finite value to a small value beyond cutoff. A value of δ equal to 0.2 is used.

2.11 Summary

In this chapter, we have derived an analytic model of the MESFET based on a two-region description of the channel in which velocity saturation occurs in a region near the drain. The model expresses the terminal currents in terms of the terminal voltages and their rate of change, the former accounting for the DC characteristics and the latter the capacitive effects of the device. Both the drain current and the gate current must be computed from solution of nonlinear equations. The capacitive currents are derived from the idea of "terminal charge" of Ward [9].

For efficient circuit simulation and hand analysis, a simple, empirical model, such as the Statz model, is preferred.

We have not considered subthreshold current. An empirical model has been proposed [10]. Back-gating effects caused by charge traps in the substrate under high field [11] will not be discussed. Other analytic models of the MESFET based on a different velocity-field relation [8,12], or on a more complete description of the conduction process [13], or on a three-section depletion region [14], exist. Interested readers are referred to the literature for further study.

References

[1] S. Sze, *Physics of Semiconductor Devices*. Wiley Interscience, 1981. Chapter 5.

[2] A. B. Grebene and S. K. Ghandhi, "General theory for pinched operation of the junction-gate FET," Solid State Electronics, Vol. 12, 1969, pp. 573-89.

[3] E. N. Trofimenkoff, "Field-dependent mobility analysis of the FET," Proceedings of the IEEE, Vol. 53, no. 11, November 1965, pp. 1765-66.

[4] R. A. Pucel, H. A. Haus, and H. Statz, "Signal and noise properties of GaAs microwave field effect transistors," *Advances in Electronics and Electron Physics*, L. Marton, Ed., vol. 38. New York: Academic, 1975, pp. 195-265.

[5] A. Chandra, private communication.

[6]   N. G. Alexopoulo, J. A. Maupin, P. T. Greiling, "Determination of the electrode capacitance matrix for GaAs FETs," IEEE Transactions on Microwave Theory and Techniques, Vol. MTT-28, No. 5, May 1980, pp. 459-466.

[7]   H. Statz, P. Newman, I.W. Smith, R. A. Pucel, and H. A. Haus, "GaAs FET device and circuit simulation in SPICE," IEEE Transactions on Electron Devices, Vol. ED-34, No. 2, February 1987, pp. 160-169.

[8]   M. Shur, *GaAs Devices and Circuits*. New York: Plenum Press, 1987, p. 318.

[9]   D. E. Ward, "Charge-based modelling of capacitance in MOS transistors," Ph.D. dissertation, Integrated Circuits Laboratory, Stanford University, Technical Report G201-11, June 1981.

[10] J. M. Golio, J. R. Hauser, and P. A. Blakey, "A large-signal GaAs MESFET Model implemented on SPICE," IEEE Circuits and Devices Magazine, September 1985, pp. 21-29.

[11] S. J. Lee, C. P. Lee, E. Shen, and G.R. Kaelin, "Modeling of back-gating effects on GaAs digital integrated circuits," IEEE Journal of Solid State Circuits, April 1984, pp. 245-50.

[12] C-S Chang and D-Y Day, "Analytic theory for current-voltage characteristics and field distribution of GaAs MESFETs," IEEE Transactions on Electron Devices, Vol. 36, No. 2, February 1989, pp. 269-80.

[13] Y-K Feng and A. Hintz, "Simulation of submicron GaAs MESFET's using a full dynamic transport model," IEEE Transactions on Electron Devices, Vol 35, No. 9, September 1988, pp. 1419-31.

[14] H. C. Ki, S. H. Son, K Park, and K. D. Kwack, "A three-section model for computing I-V characteristics of GaAs MESFET's," IEEE Transactions on Electron Devices, Vol. 34, No. 9, September 1987, pp. 1929-33.

Chapter 3

Enhancement-Depletion Logic Circuits

3.1 Introduction

In the next four chapters, we shall study the design of digital circuits made of GaAs MESFETs. We recall from Chapter 1 that the low-field mobility and peak electron velocity in GaAs are each several times greater than their respective values in Si. Also, in a GaAs device, there is no ground plane and all capacitances between conductors and device terminals are lateral. Moreover, there is no junction capacitance between the drain or source and the substrate as in MOS devices. We expect therefore that the operating speeds of a GaAs transistor to be at least several times faster than those of an MOS transistor of comparable gate length and width. Our task at hand is to look for ways to harness the speed advantage of the GaAs transistor in order to design circuits that will operate at bit rates of several Gigabits per second, or higher. Such circuits are used in fiber-optic communications systems, supercomputers, and real-time signal processing systems.

The current-voltage characteristics of a GaAs MESFET are similar to those of an NMOS transistor. This suggests that many of the design techniques that have been developed for NMOS digital circuits can be applied directly to the design of GaAs digital circuits. However, there are fundamental differences between a GaAs transistor and an MOS transistor. Foremost among them is that in a GaAs device, when the gate voltage exceeds the Schottky barrier voltage, the gate becomes conducting and the gate current has detrimental effects on circuit performance. In order to counter these effects, special design techniques need be developed and GaAs logic circuits are somewhat more complicated than comparable NMOS circuits. In addition, when the gate becomes conducting, its voltage is clamped at a value equal to the barrier voltage. This limits the logic swing to a fraction of the supply voltage and places severe requirements on the noise margins of logic circuits. Other differences will be apparent as we introduce the logic families.

We shall first simplify the current-voltage characteristics of the GaAs MESFET in order to do approximate analysis of the circuits. We then study the design of several types of the basic logic element, the inverter, with each type leading to a different family of logic circuits. Their performance will be compared with respect to noise margins, pull-up and pull-down delays, and fan-out capability. Since the accurate current-voltage and capacitance-voltage relations of a GaAs transistor are so complex, it is not possible to derive analytic expressions of these performance measures. Computer simulation will be necessary but graphical solutions and approximate analytic solutions will be used whenever practical, in order to gain a quantitative understanding of the relationship between device characteristics and circuit performance.

3.2 Simplified Device Model for Circuit Design

From the last chapter, we saw that the drain current, gate current and terminal charges are complicated functions of the drain and gate voltages. When velocity saturation is taken into consideration, as we must, the functions can only be evaluated by iteration. While such an accurate model of the device is necessary for computer simulation, it is hardly useful for hand analysis, which is required to guide the design of circuits. For the latter purpose, a simplified model such as that of Fig. 2.10.4 is sufficient. Furthermore, whenever a first order estimate of the circuit performance is called for, the series resistances will often be omitted and the gate-to-drain and gate-to-source capacitances are assumed to be linear and constant.

In this and all subsequent chapters, we shall use Statz' model for analysis and computer simulation. The transistor is assumed to have a gate length $L=1\mu M$, gate width $W=1000\mu M$ and the following parameters (see Eq. 2.10.1, Eq. 2.10.12 and Fig. 2.10.4):

Parameter	Enhancement	Depletion	Unit
β	0.300	0.125	AV^{-2}
V_T	0.100	-1.00	V
b	0.300	0.300	V^{-1}
λ	0.100	0.100	V^{-1}
α	2.00	2.00	V^{-1}
R_S	0.500	0.500	Ω
R_D	0.500	0.500	Ω
R_G	0.0	0.0	Ω
$*I_S$	1.0×10^{-14}	1.0×10^{-14}	A
C_{gso}	0.4×10^{-12}	0.4×10^{-12}	F
C_{gds}	0.4×10^{-12}	0.4×10^{-12}	F
V_{bi}	1.0	1.0	V

* Saturation current of gate-source and gate-drain diodes.

3.3 E-D Logic

The first logic family to be studied is the enhancement-depletion (E-D) logic. Its inverter circuit is shown in Fig. 3.3.1. The enhancement transistor T_e is sometimes called the pull-down or driver transistor, and the depletion transistor T_d is called the pull-up or load transistor. Since the threshold voltage of T_d is negative, with its gate connected to the source, T_d is always conducting and it can be regarded simply as a non-linear resistor. In fact, in fabricating the E-D pair, the depletion transistor is often ungated, i.e., its gate is left unmade. Its current-voltage relation is similar to that of a gated depletion transistor and can be approximated by a tanh function [1].

An inverter is rarely unterminated. Its output is usually connected to the gate of one or more transistors (fan-out). For simplicity, we shall represent the input circuit of a fan-out transistor by a linear capacitor and a gate-to-source diode, its series source resistance being neglected for now. The gate-to-drain diode can be omitted since, as we will see, the gate voltage is limited by the diode to about 0.8 V and the drain voltage of the fan-out transistor is about 0.2-0.3 V with respect to the source when its gate voltage is at the highest value, so the gate-to-drain diode current is negligible.

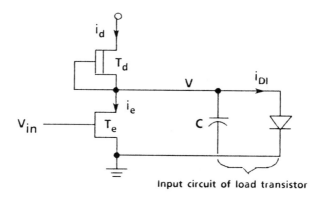

Input circuit of load transistor

Fig. 3.3.1 An enhancement-depletion inverter. Its load
is represented by a capacitor in parallel with a diode.

Graphical Solution

In Fig. 3.3.2(a), (b), and (c) we show the currents I_d, I_e and I_{DI} each
as a function of the output voltage V, respectively, with the quantities
defined in Fig. 3.3.1. At DC, the circuit equation is

$$I_d = I_e + I_{DI} \tag{3.3.1}$$

Combining the set of curves of I_e with the diode current, we get the
curves shown in Fig. 3.3.3. Superimposing the curve of I_d on them, we
obtain the solution of the circuit as intersections at points A, B, C, and D
for input voltages 0.2, 0.4, 0.6 and 0.8 V, respectively. Plotting the output voltages V at these points versus the input voltage V_{in}, we get the
transfer characteristics of the inverter shown in Fig. 3.3.4.

When the input voltage exceeds 0.8 V, the gate current of T_e begins
to flow. Since the drain of T_e is low at the same time, its gate-to-source
and gate-to-drain diodes are both conducting. The gate-to-drain diode
current I_{GDD} flows in a direction opposite to that of the drain current and
the circuit equation becomes

$$I_d = I_e - I_{GDD} + I_{DI} \tag{3.3.2}$$

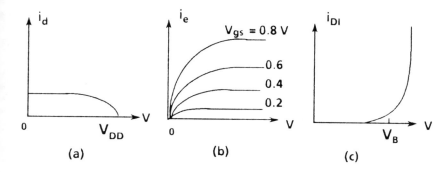

Fig. 3.3.2 (a) Depletion current, (b) enhancement current, and (c) diode current, each as a function of the output voltage v.

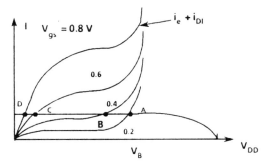

Fig. 3.3.3 I-V plots associated with an ED inverter.

I_d is reduced and the voltage drop across T_d is smaller. So the output voltage V will rise. Fig. 3.3.5 shows the situation when V_{in} is greater than 0.8 V with the solution points indicated as E and F, which are also shown in the transfer characteristics of Fig. 3.3.4.

Comparison with NMOS Inverter

It is now evident that the transfer characteristics of a MESFET E-D inverter are different from those of an NMOS E-D inverter in many aspects:

(1) When the input is low so that T_e is cut off, the output voltage is not at V_{DD} but is clamped at the turn-on voltage of the gate-to-source diode of the following stage. Moreover, the current in the

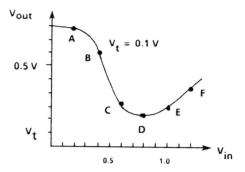

Fig. 3.3.4 Transfer characteristics of an ED inverter.

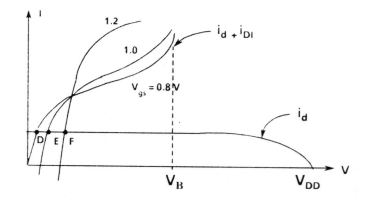

Fig. 3.3.5 I-V plots showing the region when the
gate-to-drain current is substantial.

pull-up transistor T_d is not zero and power is dissipated even when
the pull-down transistor T_e is cut off.

(2) When the input is high, the output is low up to a value
corresponding to an input of about 0.8 V. Any further increase of
the input results in an increase of the output. It is possible that this
supposedly low but actually high value of output will produce a
logical error.

(3) The voltage swing is limited to about 0.8 V, so that the E-D
inverter is a low power logic circuit. However, the small voltage
swing places severe requirements on the tolerance of the device

parameters, in particular, the threshold voltage, as we will see later.

To overcome these limitations, means must be found to prevent the onset of gate current and special techniques must be devised to make circuits less sensitive or more tolerant to threshold variations.

Noise Margins

Noise margins of an inverter are a measure of its immunity against the possibility of producing a logical error (high instead of low, and vice versa) owing to impulsive noise injected at a node, or to variation of the logical high and low levels over a chain of inverters or logic gates. They are best obtained by superimposing the transfer curves of two identical inverters with the input of one being the output of the other, as shown in Fig. 3.3.6. Noise margins can be defined in many arbitrary ways. On the transfer curve, the points at which the slope is -1 mark the two critical input values, V_{IL} (input low) and V_{IH} (input high). The minimum value of the curve is V_{OL} (output low) and the maximum is V_{OH} (output high). We define the noise margins to be:

$$NML\,(noise\ margin\ low) = V_{IL} - V_{OL} \qquad (3.3.3)$$

$$NMH\,(noise\ margin\ high) = V_{OH} - V_{IH} \qquad (3.3.4)$$

The significance of NML is that suppose inverter I is at logical low at its output, then the output voltage level can vary from its minimum value of V_{OL} to as high a value as V_{IL} and the output of inverter II will remain at logical high. Similarly for NMH.

The noise margins of an E-D inverter depends on the ratio of the gate width of the depletion transistor W_d to the gate width of the enhancement transistor W_e. Fig. 3.3.7 shows computer simulation results of the transfer curves for various values of W_d/W_e. It is seen that the inverter DC characteristics are improved as we make the depletion transistor smaller with respect to the enhancement transistor. This can also be seen graphically as shown in Fig. 3.3.8.

Table 3.3.1 shows how the noise margins depend on W_d/W_e.

For a given threshold voltage V_T, the noise margins improve somewhat as we decrease the ratio W_d/W_e. What is more noteworthy is how the noise margins vary with the threshold voltage. For a fixed W_d/W_e,

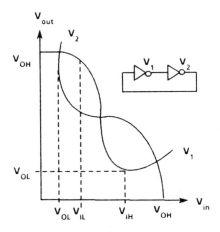

Fig. 3.3.6 Definition of noise margins.

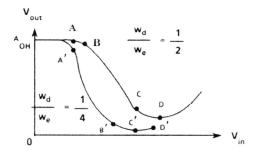

Fig. 3.3.7 Transfer curves of an E-D inverter for
different W_d/W_e.

say 1/2, NML increases from 0.15 to 0.41 V whereas NMH decreases
from 0.49 to 0.22 V. Since logic signals usually propagate along a chain
of inverters, the two noise margins should have about equal value. The
optimum choice of W_d/W_e and V_t seems to be, for this particular device,
1/2 - 1/3, and 0.2, respectively.

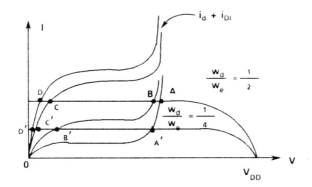

Fig. 3.3.8 Solution points A, B, C, and D shift to A', B', C' and D' as
W_d/W_e is reduced.

V_T	W_d/W_e	NML	NMH
0.0	1/1	0	0
0.0	1/2	0.15	0.49
0.0	1/3	0.15	0.50
0.1	1/1	0	0
0.1	1/2	0.25	0.37
0.1	1/3	0.25	0.41
0.2	1/1	0	0
0.2	1/2	0.32	0.28
0.2	1/3	0.33	0.34
0.3	1/1	0	0
0.3	1/2	0.41	0.22
0.3	1/3	0.43	0.22

Table 3.3.1   Noise margins as functions of the threshold voltage and
width ratio.

Threshold Variation

Another measure of sensitivity of circuit performance to threshold
variation is to determine how the transfer curve will shift as the threshold
voltage changes.

With reference to Fig. 3.3.9 in which we have shown the transfer
curves of two consecutive inverters in a chain, let V_{ct} be the input voltage at which the two curves cross each other, namely, the output equals

the input. We shall refer to this voltage as the **circuit threshold** in the sense that if the input is significantly lower than V_{ct}, the output of an inverter will surely be high, and if the input is significantly higher than V_{ct}, the output will be low. In a logic circuit, we should maintain the same circuit threshold from one logic stage to another. Let us examine how variations of the device threshold voltage V_T will affect the circuit threshold.

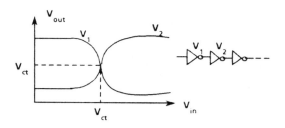

Fig. 3.3.9 Definition of circuit threshold voltage:
$V_{out} = V_{in} = V_{ct}$.

In Fig. 3.3.10, we have plotted the I-V characteristics of an ED inverter. Suppose V_T is increased by some small amount. Then the current I_e will be lowered for the same value of input, as indicated by dashed lines. The solution points will shift to the right and the transfer characteristics will also shift to the right. The new circuit threshold will be greater than its old value. Simulation results are shown in Fig. 3.3.11. It would be desirable to design circuits in which the circuit threshold is insensitive to the threshold voltage of the device. Examples of such circuits will be given in later chapters.

As a second example, computer simulation of a ring oscillator consisting of seven stages of ED inverter shows that sustained oscillation is possible for threshold voltage equal to 0, 0.1, 0.2 V. When it exceeds 0.3 V, oscillation stops, as shown in Fig. 3.3.12.

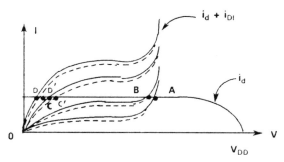

Fig. 3.3.10 I-V characteristics when the threshold
voltage is increased (shown in dashed line).

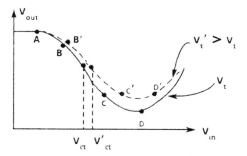

Fig. 3.3.11 Circuit threshold shifts as V_t varies.

3.4 Noise Margin Analysis

While graphical solution gives us a qualitative description of the circuit performance and computer simulation confirms it, neither can serve as design guidelines that provide insights into design trade-offs. What is needed are analytic formulas for the performance measures such as noise margins and pull-up and pull-down delays.

Since the circuit is nonlinear, it is impossible to obtain closed form expressions for these measures. However, approximations can be made that will simplify the analysis to the extent that the noise margins and delays can be computed easily in terms of the device parameters without simulation.

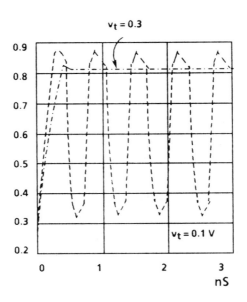

Fig. 3.3.12 Simulation output of a 7-stage ring oscillator showing that
oscillation fails at $V_t \geq 0.3V$.
voltage is increased (shown in dashed line).

Consider Fig. 3.4.1, in which we show an ED inverter loaded by
another. We assume an idealized transfer characteristics with critical
voltages defined as shown. In this section, we will derive expressions for
the low noise margin $NML=V_{IL}-V_{OL}$ and high noise margin
$NMH=V_{OH}-V_{IH}$ in terms of the device characteristics.

To Find V_{OH} and V_{OL}

With reference to Fig. 3.4.1(a), consider the case where $V_{in}=V_{OL}$
and $V=V_{OH}$. If V_{OL} is sufficiently low, V_{OH} will be high enough to cause
gate current to flow in the second stage. Moreover, we expect the drain
current of the second stage will be substantial so that the voltage drops
across the series source and drain resistances become significant. Fig.
3.4.2 shows the equivalent circuit that we must analyze. In a properly
designed circuit, the output of the second stage must be at least as low as
V_{OL} in order that the pair of inverters be "self- restoring." Let the voltages and currents be as defined in the figure and we denote the drain
current of a transistor by the convention $I(V_D, V_G, V_S)$, where

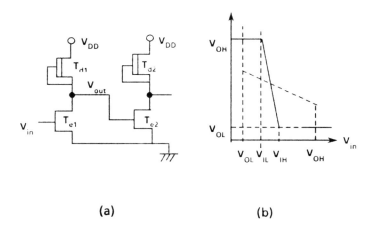

Fig. 3.4.1 (a) Loaded E-D inverter. (b) Idealized transfer
curve and definition of critical voltages.

$V_D, V_G,$ and V_S are the drain, gate, and source voltages, respectively with
respect to some reference such as ground. Writing circuit equations, we
have

$$I_{d1}(V_{DD},V_{OH},V_{OH})=I_{e1}(V_{OH},V_{OL},0) + I_{DI} \tag{3.4.1}$$

$$V_{OH}=V_{DI} + V_s \tag{3.4.2}$$

$$V_S=R_S\left[I_{DI} + I_{d2}(V_{DD},V_{OL},V_{OL})\right] \tag{3.4.3}$$

$$I_{DI}=I_s(exp\,(qV_{DI}/nkT) - 1) \tag{3.4.4}$$

where I_s is the diode saturation current and n the ideality constant. Combining equations, we get

$$V_{OH} = \frac{nkT}{q}\log\frac{I_{d1}(V_{DD},V_{OH},V_{OH}) - I_{e1}(V_{OH},V_{OL},0)}{I_s} \tag{3.4.5}$$

$$+ R_s\left[I_{d1}(V_{DD},V_{OH},V_{OH}) - I_{e1}(V_{OH},V_{OL},0) + I_{d2}(V_{DD},V_{OL},V_{OL})\right]$$

Eq.(3.4.5) is implicit in the unknowns V_{OH} and is one of the two relations between V_{OH} and V_{OL}. The second relation is obtained by equat-
ing the drain currents of T_{e2} and T_{d2}, i.e.,

$$I_{d2}(V_{DD},V_{OL},V_{OL})=I_{e2}(V_{OL},V_{OH},V_S) \tag{3.4.6}$$

with V_S given by

$$V_S = R_S \left[I_{d1}(V_{DD}, V_{OH}, V_{OH}) - I_{e1}(V_{OH}, V_{OL}, 0) \right. \tag{3.4.7}$$

$$\left. + I_{d2}(V_{DD}, V_{OL}, V_{OL}) \right]$$

Eqs.(3.4.5-7) constitute a set of three nonlinear equations and three unknowns V_{OH}, V_{OL}, and V_S and can be solved by Newton iteration. However, an examination of the I-V characteristics shows that

$$I_{d2}(V_{DD}, V_{OL}, V_{OL}) \approx I_{d2(sat)} \equiv \beta_{d2} \frac{(-V_{td})^2}{1 - b_d V_{td}} \tag{3.4.8}$$

which is the saturated value of the drain current of T_{d2}, based on the Statz model. (See Eq. 2.10.1 for definition of the parameters. The subscript d signifies "depletion" and e "enhancement.") In addition, the solution point H will not change much if we ignore the drain current of T_{e1}, namely we find V_{OH} by setting $I_{e1}(V_{OH}, V_{OL}, 0)$ to zero. Accordingly, V_{OH} is given by

$$V_{OH} = \frac{nkT}{q} \log \frac{I_{d1}(V_{DD}, V_{OH}, V_{OH})}{I_s}$$

$$+ R_s \left[I_{d1}(V_{DD}, V_{OH}, V_{OH}) + I_{d2(sat)} \right] \tag{3.4.9}$$

which can be solved by iteration with the first iterate found by replacing $I_{d1}(V_{DD}, V_{OH}, V_{OH})$ by

$$I_{d1(sat)} = \beta_d \frac{(-V_{td})^2}{1 - b_d V_{td}} \tag{3.4.10}$$

To find V_{OL}, we need to express the currents explicitly. Doing so for Eq. (3.4.6), we get

$$I_{d2(sat)} = \beta_{e2} \frac{(V_{OH} - V_S - V_{te})^2}{1 + b_e(V_{OH} - V_S - V_{te})} \tag{3.4.11}$$

$$\tanh \alpha_e(V_{OL} - V_S - R_D I_{d2(sat)}) \left[1 + \lambda_e(V_{OL} - V_S - R_D I_{d2(sat)}) \right]$$

Since V_{OL} will be small, the last term can be approximated by unity and the equation can be solved to give

$$V_{OL} = V_S + R_D I_{d2(sat)} \tag{3.4.12}$$

$$+ \frac{1}{2\alpha_e} \log \frac{\beta_e (V_{OH} - V_S - V_{te})^2 / I_{d2(sat)} + \left[1 + b_e (V_{OH} - V_S - V_{te}) \right]}{\beta_e (V_{OH} - V_S - V_{te})^2 / I_{d2(sat)} - \left[1 + b_e (V_{OH} - V_S - V_{te}) \right]}$$

with

$$V_S = R_S \left[I_{d1}(V_{DD}, V_{OH}, V_{OH}) + I_{d2(sat)} \right] \tag{3.4.13}$$

and V_{OH} given by Eq. (3.4.9).

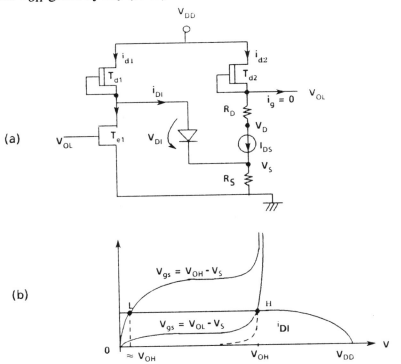

Fig. 3.4.2 (a) Equivalent circuit for the computation of V_{OH} and V_{OL}. (b) I-V plots showing solution points H and L.

With reference to Eq. (3.4.5), we see that in order to make V_{OH} large as desired in a good inverter design, we should make the ratio of the depletion to enhancement current, or W_d/W_e large. However, any increase in this ratio will increase V_{OH} only logarithmically. Similarly, from Eq. (3.4.12), V_{OL} can be lowered by making W_d/W_e small with logarithmic improvement. In an actual circuit, V_{OH} is essentially fixed by the Schottky barrier voltage. The optimal value of this ratio will dependent on other considerations such as the pull-up delay, to be discussed later.

As a corollary of these observations, we see from Eqs. (3.4.5) and (3.4.12) that V_{OH} and V_{OL} cannot be arbitrarily set as design objective. In other words, for a given pair of values of V_{OH} and V_{OL}, we may not be able to find acceptable ratios W_{d1}/W_{e1} and W_{d2}/W_{e2} that satisfy these two equations. Consequently we can state that

> An inverter chain is "self-restoring" if and only if for any adjacent pair of inverters, a solution exists for the pair of equations (3.4.5) and (3.4.12).

To Find V_{IL}

With reference to Fig. 3.4.1, we shall interpret V_{IL} as the input voltage such that if the input exceeds this value by a slight amount the output will drop to a value just low enough to cause the diode current of T_{e2} to become zero. Fig. 3.4.3 shows the equivalent circuit and the I-V characteristics. The circuit equation is

$$I_{d1}(V_{DD},V_{OH},V_{OH})=I_{e1}(V_{OH},V_{IL},0) \qquad (3.4.14)$$

Since V_{OH} has already been computed and is known, we solve for V_{IL} as given below:

$$V_{IL}=V_{te}+V_{ch}\sqrt{1+V_{ch}^2 b_e^2/4}+V_{ch}^2 b_e/2 \qquad (3.4.15)$$

where

$$V_{ch}^2 = \frac{I_{d1}(V_{DD},V_{OH},V_{OH})}{\beta_{e1}(1+\lambda_e V_{OH})\tanh\alpha_e V_{OH}}$$

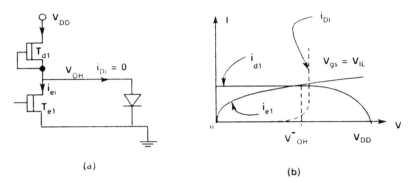

Fig. 3.4.3 (a) Equivalent circuit to compute V_{IL}. (b) Graphical solution.

To Find V_{IH}

We interpret V_{IH} to be the largest input voltage for which there is no gate current in the enhancement transistor T_{e1} and its output voltage equals V_{OL}. Following the same derivation as V_{IL}, we find

$$V_{IH} = V_{te} + V_{cl}\sqrt{1+V_{cl}^2b_e^2/4} + V_{cl}^2b_e/2 \qquad (3.4.16)$$

where

$$V_{cl}^2 = \frac{I_{d1}(V_{DD},V_{OL},V_{OL})}{\beta_{e1}(1+\lambda_e V_{OL})\tanh\alpha_e V_{OL}}$$

Noise Margins

We now have $V_{OH}, V_{OL}, V_{IL},$ and V_{IH}. The noise margins are given by

$$NMH = V_{OH} - V_{IH} \qquad (3.4.17)$$

$$NML = V_{IL} - V_{OL} \qquad (3.4.18)$$

Fig. 3.4.4 shows the computed noise margins, each as a function of the ratio $W_{d1}/W_{e1} = W_{d2}/W_{e2}$ for various values of the threshold voltage. It is seen that to obtain a good inverter design so that NML = NMH, we would choose, for this example, a ratio of 1/2 with $V_{te} = 0.1V$.

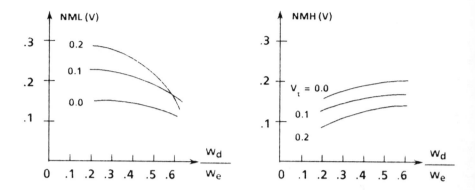

Fig. 3.4.4 Noise margins as a function of width ratio W_d/W_e
with threshold voltage of the enhancement transistor as a parameter.

3.5 Pull-Up Delay

In most studies of digital circuits, pull-up and pull-down delays are derived on the basis of a step input. Pull-up delay is defined as the time elapsed when the output voltage reaches some fraction of its steady state value; similarly for pull-down delay. The delay of a logic gate is taken to be the average of the two. Over a chain of inverters, the delay is the sum of the delays of the individual gates. This is clearly incorrect since the inputs to the intermediate inverters are not a step but rather a pulse with finite rise and fall times. The computed delay is therefore overly optimistic.

In this section, we will derive the pull-up and pull-down delays of an inverter due to an input pulse with finite rise and fall times. To get a closed form expression, approximations will be made and these will be clearly stated.

Consider Fig. 3.5.1. We assume the inverter is terminated by another and that the input circuit of the second stage can be represented by a diode in parallel with a linear capacitor. This is of course a gross approximation. However, if the voltage swing is not so large, the estimate of the delay on this basis is acceptable.

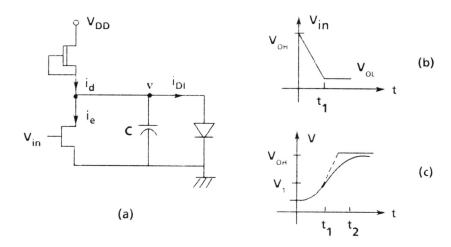

Fig. 3.5.1 Computation of pull-up delay. (a) The equivalent
circuit; (b) input waveform; and (c) output waveform.

The input waveform is shown in the figure. It falls from V_{OH} to
V_{OL} linearly in t_1 seconds. During this time, we assume the output voltage will rise from V_{OL} to V_{OH} and is clamped at V_{OH} by the gate-to-
source diode of the following stage. In the I-V plane, shown in Fig.
3.5.2, we have plotted the depletion current I_d, the enhancement current
I_e and the diode current I_{DI}. The solution trajectory is shown in a dashed
line. Let

$$V_1 = V(t_1) \quad \text{and} \quad V(t_2) = V_{OH}$$

We wish to find t_2, which will be defined as the pull-up delay.

The circuit equation is

$$C\frac{dV}{dt} = I_d(V_{DD}, V, V) - I_e(V, V_{in}, 0) - I_{DI} \qquad (3.5.1)$$

For the period $0 \leq t < t_2$, I_{DI} will be small and it will be neglected. So t_2
can be interpreted as the time at which the diode just begins to conduct.
Integrating Eq.(3.5.1) and approximating the integrals over the two time
intervals $[0, t_1]$ and $[t_1, t_2]$ by the trapezoidal rule, we get

$$t_{pu} = \frac{2C(V_{OH} - V_{OL}) + t_1\left[I_d(V_{DD}, V_{OH}, V_{OH}) - I_e(V_{OH}, V_{OL}, 0)\right]}{I_d(V_{DD}, V_{OH}, V_{OH}) - I_e(V_{OH}, V_{OL}, 0) + I_d(V_{DD}, V_1, V_1) - I_e(V_1, V_{OL}, 0)}$$

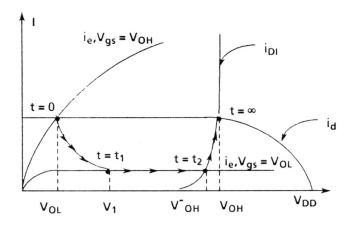

Fig. 3.5.2 Solution trajectory of pull-up delay on the I-V plane.

$$(3.5.2)$$

It should be noted that the first two terms in the denominator represent the capacitor current at $t=t_2$ and the last two terms represent the capacitor current at $t=t_1$. The pull-up delay has two components, one being proportional to C and to the voltage swing $V_{OH}-V_{OL}$, and the second being due to the finite fall time of the input. If the input is a step, $t_1=0$ and $V_1=V_{OL}$. The pull-up delay becomes

$$t_{pu} = \frac{2C\,(V_{OH}-V_{OL})}{\left[I_d(V_{DD},V_{OH},V_{OH}) - I_e(V_{OH},V_{OL},0) + I_d(V_{DD},V_{OL},V_{OL})\right.} \\ \left. - I_e(V_{OL},V_{OL},0\right]} \qquad (3.5.3)$$

as a special case. It remains to find V_1. Integrating Eq.(3.5.1) from $V=V_{OL}$ to $V=V_1$ over dV and from $t=0$ to $t=t_1$ over dt, and solving for V_1, we get

$$V_1 = V_{OL} + \frac{t_1}{2C}\left[I_d(V_{DD},V_1,V_1)-I_e(V_1,V_{OL},0\right] \qquad (3.5.4)$$

which is implicit in V_1 and must be solved by iteration with an initial guess, say $1.1V_{OL}$.

From Eq. (3.5.3), we see that to reduce the pull-up delay, we must make the ratio W_d/W_e large. This is confirmed in Fig. 3.5.3 in which we show the computed results of the delay as a function of W_d/W_e with the threshold voltage of the enhancement transistor as a parameter, for both a step input and an input with finite fall-time. It is seen that the pull-up delay is not a very sensitive function of the threshold voltage. However, the delay is reduced with a larger threshold voltage. This can be seen by noting that a larger threshold will result in a smaller voltage swing $V_{OH}-V_{OL}$. On the other hand, the high noise margin will be reduced as the threshold voltage is increased. Some trade-off will have to be made. In Fig. 3.5.4, we show the pull-up delay as a function of the fall-time of the input. It should be specially noted that the pull-up delay can be smaller than the fall-time.

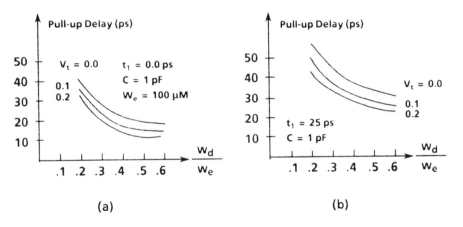

Fig. 3.5.3 Pull-up delay of a loaded ED inverter. (a) Input is a step. (b) Input is a pulse with a fall-time of 25 ps.

3.6 Pull-Down Delay

Fig. 3.6.1 shows the circuit from which the pull-down delay is computed. The input rises linearly from V_{OL} to V_{OH} in time t_1. For $t \leq 0$, $V=V_{OH}$ and the gate-to-source diode of the following stage is conducting. In fact, it remains conducting until the input rises to some value V_{IL} at time t_0. The output voltage falls to V_{OL} at time t_2, passing through V_0

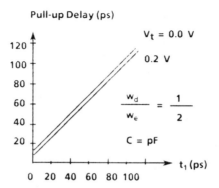

Fig. 3.5.4 Pull-up delay as a function of the fall-time of the input.

at t_0 and V_1 at t_1, as shown in Fig. 3.6.1(c). Correspondingly, the diode current drops from its initial value to zero at t_0, while the current I_e rises from its initial value to a maximum at t_1 and falls to a value equal to I_d at t_2. This is depicted in Fig. 3.6.1(d). The difference of I_e and I_d is the discharge current of the capacitor. The solution trajectory can be seen in the I-V plane, shown in Fig. 3.6.2.

To find the pull-down delay, we divide the time of interest into three intervals.

(1) $0 \leq t < t_0$. Here we have $V_{OH} \geq V > V_0$ and $V_{OL} \leq V_{in} < V_{IL}$. Integrating the circuit equation as usual and noting that at $t=0$, $I_d - I_e - I_{DI} = 0$, and that at $t=t_0$, $I_{DI}=0$, we get

$$V_0 = V_{OH} + \frac{t_0}{2C}\left[I_d(V_{DD},V_0,V_0) - I_e(V_0,V_{IL},0)\right] \qquad (3.6.1)$$

which is implicit in the unknown V_0 and must be solved by iteration with $V_0 = V_{OH}$ as an initial guess.

(2) $t_0 \leq t < t_1$. In this interval, $I_{DI}=0$ and $V_0 \geq V > V_1$ and $V_{IL} \leq V_{in} < V_{OH}$. Moreover, the enhancement transistor current I_e will be substantial and the voltage drops across the series drain and source resistances should be included. Let

$$V_d = V_1 - R_D I_e(V_d,V_{OH},V_s) \qquad (3.6.2)$$

$$V_s = R_S I_e(V_d,V_{OH},V_s) \qquad (3.6.3)$$

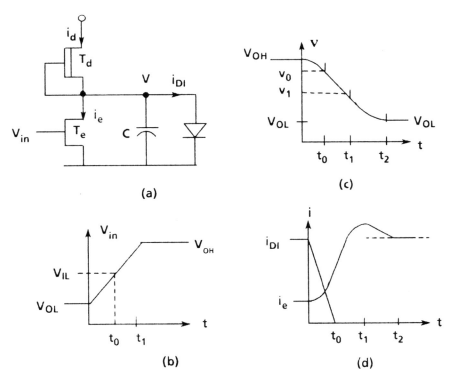

Fig. 3.6.1 (a) Circuit from which pull-down delay is computed.
(b) Input waveform. (c) Output voltage. (d) Current waveforms.

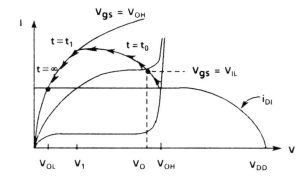

Fig. 3.6.2 Solution trajectory of pull-down delay.

Integrating the circuit equation, we get

$$V_1 = V_0 + \frac{t_1 - t_0}{2C}\left[I_d(V_{DD}, V_1, V_1) - I_e(V_d, V_{OH}, V_s)\right.$$

$$\left. + I_d(V_{DD}, V_0, V_0) - I_e(V_0, V_{IL}, 0)\right] \qquad (3.6.4)$$

The three equations must be solved simultaneously by iteration to obtain $V_1, V_d,$ and V_s. A good set of initial guess would be $V_1 = 0.9V_0$, and $V_d = V_s = 0$.

(3) $t_1 \leq t \leq t_2$. Here we have $V_1 \geq V \geq V_{OL}$, $V_{in} = V_{OH}$, and $I_{DI} = 0$. In this case, the gate-to-source diode of the enhancement transistor becomes conducting and the equivalent circuit is as shown in Fig. 3.6.3 with the voltages and currents as defined.

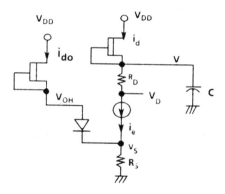

Fig. 3.6.3 Equivalent circuit for the computation of pull-down delay.

We assume the diode current is equal to the depletion current of the previous stage. So, writing circuit equations, we have

$$V_s = R_S\left[I_{d0}(V_{DD}, V_{OH}, V_{OH}) + I_e(V_d, V_{OH}, V_s)\right] \qquad (3.6.5)$$

$$V_d = V_{OL} - R_D I_e(V_d, V_{OH}, V_s) \qquad (3.6.6)$$

These equations are solved simultaneously for V_s and V_d using their values found at t_1 as initial guesses. Once V_s and V_d are found, the pull-down delay can be derived in the same way as the pull-up delay and it is given by

$$t_{pd} = t_1 + \cfrac{2C(V_1 - V_{OL})}{\begin{aligned}&\left[I_e(V_d, V_{OH}, V_s) - I_d(V_{DD}, V_{OL}, V_{OL}) + I_e(V_{d1}, V_{OH}, V_{s1})\right.\\[2mm]&\left.\qquad - I_d(V_{DD}, V_1, V_1)\right]\end{aligned}}$$

$$(3.6.7)$$

where V_{d1} and V_{s1} are the values of V_d and V_s found from Eqs. (3.6.3) and (3.6.4), respectively.

Again, for the special case of a step input, $t_1=0$ and $V_1=V_{OH}$ and

$$t_{pd} = \cfrac{2C(V_{OH} - V_{OL})}{\begin{aligned}&\left[I_e(V_{dL}, V_{OH}, V_{sL}) - I_d(V_{DD}, V_{OL}, V_{OL}) + I_e(V_{dH}, V_{OH}, V_{sH})\right.\\[2mm]&\left.\qquad - I_d(V_{DD}, V_{OH}, V_{OH})\right]\end{aligned}}$$

$$(3.6.8)$$

where V_{dH}, V_{sH}, V_{dL}, and V_{sL} satisfy

$$V_{dH} = V_{OH} - R_D I_e(V_{dH}, V_{OH}, V_{sH}) \qquad (3.6.9)$$

$$V_{sH} = R_S \left[I_e(V_{dH}, V_{OH}, V_{sH}) + I_{d0}(V_{DD}, V_{OH}, V_{OH})\right] \qquad (3.6.10)$$

$$V_{dL} = V_{OL} - R_D I_e(V_{dL}, V_{OH}, V_{sL}) \qquad (3.6.11)$$

$$V_{sL} = R_S \left[I_e(V_{dL}, V_{OH}, V_{sL}) - I_{d0}(V_{DD}, V_{OH}, V_{OH})\right] \qquad (3.6.12)$$

From Eq. (3.6.7), we see that in order to reduce the pull-down delay, we must make the ratio W_d/W_e small. However, reducing this ratio results in an increase of the voltage swing so that the delay remains essentially unchanged. This is seen in Fig. 3.6.4 where we have plotted the computed delay as a function of the width ratio with the threshold voltage of the enhancement transistor as a parameter. The effect of finite rise time of the input can be seen in Fig. 3.6.5. The delay is approximately proportional to the rise time.

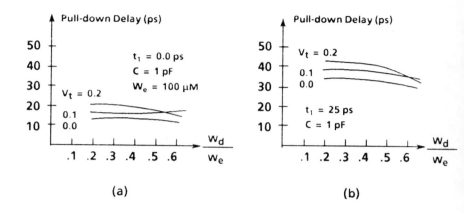

Fig. 3.6.4 Pull-down delay as a function of the width ratio. (a) Input is a step. (b) Input is a pulse with a rise time of 25 ps.

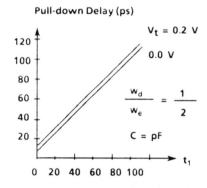

Fig. 3.6.5 Pull-down delay as a function of the rise time of the input pulse.

3.7 Fan-Out and Fan-In

The expressions of pull-up and pull-down delays are derived for the case of fan-out equal to one and fan-in equal to one. To compute the delay for the case of fan-out equal to FO and fan-in to FI (NOR gate, really), we simply replace C by $C{\times}FO$ and I_e by $FI{\times}I_e$ in these expressions. Fig. 3.7.1 shows the simulation results and the delay as a function of fan-out (FO) with the width ratio W_d/W_e and the threshold voltage as parameters.

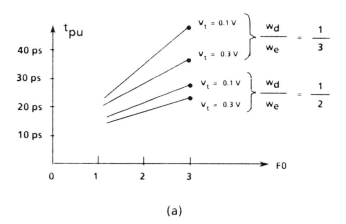

(a)

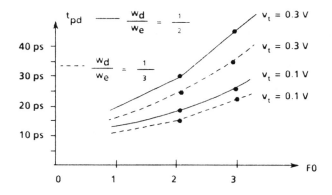

Fig. 3.7.1 (a) Pull-up delay (t_{pu}) and pull-down delay (t_{pd})
as functions of fan-put (FO) for various values of W_d/W_e and
the threshold voltage of the enhancement transistor (V_t).

The effect of fan-out on the pull-up and pull-down delays can also
be appreciated from Fig. 3.7.2 in which we show the pulse response of a
loaded ED-inverter for different values of fan-out.

3.8 NOR, NAND

An ED NOR gate is simply constructed by connecting identical
enhancement transistors in parallel to replace the pull-down transistor as
shown in Fig. 3.8.1. Let the number of inputs be FI. Then on pull-down,

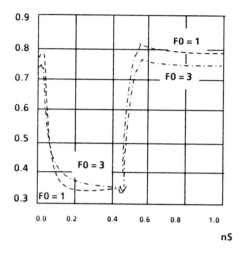

Fig. 3.7.2 Pulse response of an ED inverter obtained by computer
simulation.

the most favorable situation is for all inputs to be high so that the transistors present FI conducting paths to discharge the capacitor C. The pull-down delay will be smaller than the case of a simple ED inverter and can be computed as suggested in the last section.

On pull-up, all the inputs must be low. Since the enhancement transistors are not completely cut-off, they draw current away from the charging current supplied by the depletion transistor so that the pull-up delay will increase. To compensate for the increase, the depletion transistor should be made larger, but this will increase the pull-down delay. We see once again that trade-offs between pull-up and pull-down delays are possible. In practice, in order not to increase the pull-up delay excessively, the number of inputs rarely exceeds three.

As to the NAND gate, shown in Fig. 3.8.2, we recall that in an ED inverter, when the input is high, the output V_{OL} often exceeds the threshold voltage V_{te}. When two enhancement transistors are connected in series as in an NAND gate, the output voltage will be significantly higher than V_{te} so that the following stage will probably have an output significantly lower than V_{OH}, and we have a logical error. For this reason, NAND gates are seldom used in GaAs circuits.

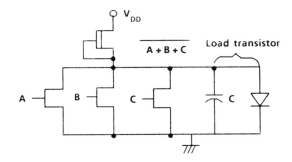

Fig. 3.8.1 An ED NOR gate.

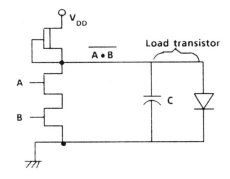

Fig. 3.8.2 An ED NAND gate.

3.9 Flip Flops

ED flip-flops are best constructed out of NOR gates. A 6-NOR D-flip-flop is shown in Fig. 3.9.1. As long as CK is high, nodes A and B will be low and the output latch is undisturbed, i.e., its outputs Q and $\overline{Q}$ remains unchanged, regardless of the value of DATA. When CK changes from high to low, then one of the two input latches must change. For example, if DATA is high, then node D is low. Since node A was low, node C is high, and the upper latch is unchanged. In the lower latch, node B must change to high, and Q becomes high, thus copying the DATA. Similarly, if DATA is low, node D changes to high, node C to low, node A to high, node B to low, and Q to low. When Ck changes from low to high, the output latch is effectively disconnected from the input latches and the outputs remain unchanged as explained earlier. This circuit is a "negative edge triggered" D-flip-flop.

For the circuit to operate at the highest clock speed possible, each latch must be balanced with respect to pull-down and pull-up delays. In the upper latch, NOR gate 1 has a fan-out of one whereas NOR 2 must drive three transistors. NOR gate 2 should therefore be larger to compensate for the imbalance of the two pull-up delays. Similarly, in the lower input latch, NOR gate 3 has three inputs so that its pull-down delay will be shorter than that of NOR gate 4. The enhancement transistor of NOR gate 4 should therefore be larger. Computer simulation results show that at a clock period of 400 ps and a threshold voltage of the enhancement transistors equal to 0.1 V, the circuit operates properly. The circuit fails at a clock period of 200 ps. When the threshold voltage is increased to 0.3 V, the circuit fails to operate as a flip-flop.

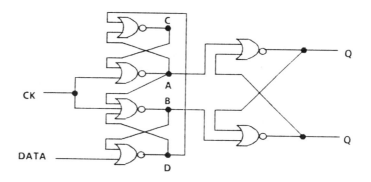

Fig. 3.9.1 A six-NOR D-flip-flop.

3.10 Remarks

The simplicity of the ED inverter makes it an attractive candidate for large scale integration. It is area and power efficient in that a logic function can be realized with very few transistors, and that when the input to an inverter is low, the inverter draws very little current. Moreover, since the logic swing is low, being about 0.6 V, the power supply voltage can be low, thus reducing power consumption. When properly designed, the ED inverter is fast because the number of logic stages over which a signal must travel is small. However, its noise margins are poor compared to other types of logic, as we will see later, and it is sensitive to threshold variation. In later chapters, we will introduce circuits that are tolerant to such variations.

References

[1]   M. Shur, *GaAs Devices and Circuits*. Plenum Press, 1987, pp. 358-367.

[2]   C. P. Lee, B. M. Welch, and R. Zucca, "Saturated resistor load for GaAs integrated circuits," IEEE Transactions on Microwave Theory and Techniques, Vol. MTT-30, No. 7, July 1082, pp. 1007-1013.

Chapter 4

Transmission-Gate Logic

4.1 Introduction

Our experience with NMOS logic circuits using transmission gates as switches suggests that we may extend the technique to GaAs circuits. However, the lack of DC isolation between the gate and the source or drain in a GaAs MESFET makes the use of the transmission gate problematic. Fig. 4.1 shows a transmission gate which is either a depletion or enhancement transistor. The circuit is symmetrical with respect to the gate, and current can flow from node 1 to node 2, or vice versa, depending on whether V_1 is greater or less than V_2. Without loss of generality, node 1 will be taken as the input and node 2 the output, which is usually connected to an inverter whose loading effect is approximated by the capacitance C. If the control signal, usually the gate voltage, is sufficiently positive with respect to either the drain or source voltage, gate current will flow and the control signal will appear at the output. In addition, if the gate-to-source or gate-to-drain capacitance is comparable to the load capacitance C, a significant fraction of the control signal will be at the output and a logical error may be produced at the output of the inverter. Nevertheless, if the amplitude of the control voltage is limited to certain range of values, the feed-through can be minimized and the transmission gate can be used as a voltage-controlled switch.

In this chapter, we will analyze the transmission gate and derive the permissible range of values that the gate voltage may assume for proper operation as a switch. In addition, we will derive expressions for the pull-up and pull-down delay when a step voltage is applied to the gate. The case of a pulse of finite rise and fall times can be similarly treated but will be omitted for brevity.

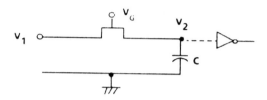

Fig. 4.1.1 GaAs transmission gate.

4.2 Analysis

First, we consider the cut-off condition. The transmission gate will not be conducting if at any time the gate voltage V_G satisfies

$$V_G \leq V_2 + V_t, \quad \text{if} \quad V_1 > V_2 \tag{4.2.1}$$

$$V_G \leq V_1 + V_t, \quad \text{if} \quad V_2 > V_1 \tag{4.2.2}$$

where V_t is the threshold voltage.

Turn-On Analysis

Now consider the situation when V_G changes suddenly from a low value V_{GL} to a high value V_{GH} sufficient to turn on the transistor. There are four cases to examine.

Case (1): $V_1 = V_H$(high) and $V_2(0) = V_L$(low) as depicted in Fig. 4.2.1 where the equivalent circuit for $t > 0$ is also shown. Writing circuit equations, we have

$$(C + C_{g2}) \frac{dV_2}{dt} = I + I_{DI} \tag{4.2.3}$$

At $t = 0^+$, the capacitor voltage must adjust itself to conserve charge and it attains a value

$$V_2(0^+) \equiv V_L^+ = V_L + \Delta V \tag{4.2.4}$$

where

$$\Delta V \equiv \frac{C_{g2}}{C+C_{g2}}(V_{GH}-V_{GL})$$

Immediately after $t=0^+$, there is a surge of current through the diode to charge the capacitor C. As the capacitor voltage rises, the diode current rapidly goes to zero at $t=t_1$ when V_2 reaches $V_{GH} - V_B$, where V_B is the barrier voltage of the Schottky junction. Thereafter, the charging current comes only from the transistor. Fig. 4.2.2(a) shows the waveform of V_2 and the solution trajectory on the I-V plane is shown in Fig. 4.2.2(b).

When V_2 reaches $V_{GH}-V_t$, the transistor is cut off and $V_{GH}-V_t$ is the maximum steady state value. If we define the pull-up delay t_{pu} as the time taken for V_2 to rise from its initial value V_L^+ to a value V_{OH}, then integration of the circuit equation gives, after approximating the integral by the trapezoidal rule as before:

$$t_{pu} = \frac{2(C+C_{g2})(V_{OH}-V_L^+)-t_1 I_t}{I(V_H,V_{GH},V_{GH}-V_B)+I(V_H,V_{GH},V_{OH})} \qquad (4.2.5)$$

where

$$I_t = \left[I_{DI}(V_{GH}-V_L^+)+I(V_H,V_{GH},V_L^+)-I(V_H,V_{GH},V_{OH}) \right]$$

$$t_1 = \frac{2(C+C_{g2})(V_{GH}-V_B-V_L^+)}{I_{DI}(V_{GH}-V_L^+)+I(V_H,V_{GH},V_L^+)+I(V_H,V_{GH},V_{GH}-V_B)}$$

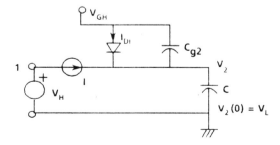

Fig. 4.2.1 Equivalent circuit to compute the pull-up delay.

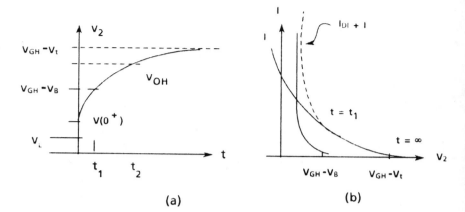

Fig. 4.2.2 (a) Output waveform in the computation of pull-up delay.
(b) Solution trajectory in the I-V plane.

Case (2): $V_2(0) = V_H$(high) and $V_1 = V_L$(low). Fig. 4.2.3 shows the situation and equivalent circuit. In this case, the capacitor will discharge through the transistor with a capacitor current equal to I until V_2 reaches $V_{GH} - V_B$ at t_1 when the gate diode begins to conduct and the discharge current is reduced by the diode current. Discharge continues until the two currents are equal and V_2 reaches its steady state value V_c. Note that the capacitor is not fully discharged because of the diode current. The waveform of V_2 and the solution trajectory on the I-V plane are given in Fig. 4.2.4.

We now define pull-down delay t_{pd} as the time at which the output voltage drops to a value V_{OL} from its initial value V_H. The circuit equation is

$$(C + C_{g2}) \frac{dV_2}{dt} = I_{DI} - I \qquad (4.2.6)$$

Integrating as usual, we get

$$t_{pd} = \frac{2(C + C_{g2})(V_H^+ - V_{OL}) - t_1 I_t}{I(V_{OL}, V_{GH}, V_L) - I_{DI}(V_{GH} - V_{OL}) + I(V_{GH} - V_B, V_{GH}, V_L)} \qquad (4.2.7)$$

where

$$I_t = \left[I_{DI}(V_{GH} - V_L^+) + I(V_H, V_{GH}, V_L^+) - I(V_H, V_{GH}, V_{OH}) \right]$$

$$t_1 = \frac{2(C+C_{g2})(V_H^+ - V_{GH} + V_B)}{I(V_H^+, V_{GH}, V_L) + I(V_{GH} - V_B, V_{GH}, V_L)} \tag{4.2.8}$$

and

$$V_H^+ = V_H + \Delta V$$

which is the initial jump of V_2 due to the step input of V_G at $t=0$.

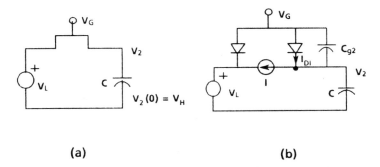

(a)   (b)

Fig. 4.2.3 Equivalent circuits for the computation of pull-down delay.

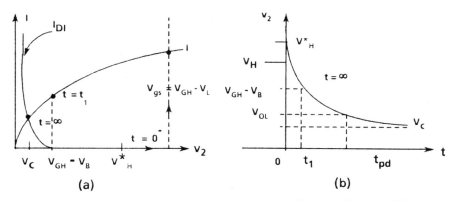

(a)   (b)

Fig. 4.2.4 (a) Output waveform in the computation of pull-down delay.
(b) Solution trajectory in the I-V plane.

Case (3): $V_1 = V_2 = V_L$(low). At $t=0^+$, V_2 jumps by an amount ΔV and decays in accordance with Case (2).

Case (4): $V_1=V_2=V_H$(high). V_2 jumps by ΔV. If $V_B>V_{GH}-V_H>V_t$ then the gate diode never conducts and V_2 decays through the transistor to a steady state value V_H. If $V_{GH}-V_H<V_t$, the transistor does not conduct and V_2 remains at $V_H+\Delta V$.

Turn-Off Analysis

For the case in which the gate voltage V_G drops suddenly from V_{GH} to V_{GL}, a value sufficient to cut off the transistor, V_2 drops by ΔV and remains at this value. If initially V_2 is high and represents a logical one, then at $t>0^+$, $V_2-\Delta V$ may be so low that it exceeds the high noise margin of the inverter and a logic error is produced at the inverter output. In a conservative design, we must limit the gate voltage swing $V_{GH}-V_{GL}$ and at the same time select an inverter with a large high noise margin as the load of the transmission gate.

From Eqs. (4.2.5) and (4.2.7), we see that to reduce the delay, we must make the transistor large. However, doing so will place a large load on the buffer that drives the gate of the transistor, and the gate voltage will rise slowly. As a result, the pull-up and pull-down delays will increase. There must therefore be a compromise.

4.3 Allowable Range of V_G

We saw in the last section that the steady value of the output is limited to a maximum value in the case of pull-up, and to a nonzero, finite value in the case of pull-down. Since the output voltage is to represent a logical 1 and 0 value, we must insure that in fact it attains these values when the transistor is turned on.

Let V_{OH} be the logical 1 value. Then in the case of pull-up (case 1), we must have, from Fig. 4.2.2(a),

$$V_{GH}-V_t>V_{OH} \quad or \quad V_{GH}>V_t+V_{OH} \tag{4.3.1}$$

Let V_{OL} be the logical 0 value. Then in the case of pull-down (case 2), the steady state value V_c will surely be lower than V_{OL} if the following is true:

$$V_{GH}-V_B<V_{OL} \quad or \quad V_{GH}<V_{OL}+V_B \tag{4.3.2}$$

Combining the two inequalities, we get a condition that V_{GH} must satisfy in order that the output represents a logical 1 and 0 correctly. The condition is

$$V_t + V_{OH} < V_{GH} < V_{OL} + V_B \qquad (4.3.3)$$

To turn off the transistor, the gate voltage V_{GL} must satisfy

$$V_{GL} < V_t + V_{OL} \qquad (4.3.4)$$

4.4 Shift Register

An obvious application of transmission gates is in the design of a dynamic shift register shown in Fig. 4.4.1, in which CK and $\overline{CK}$ are non-overlapping clock signals of period T. The circuit provides T seconds of delay on the input data signal. The clock signals can be generated by a circuit shown in Fig. 4.4.2, in which the reference signal V_{ref} must be chosen so that when CK is low, transistor T_6 conducts heavily. At the same time, when CK is high, T_6 should be cut off. The buffers must be designed so that the clock signals that drive the gates of the transmission gate transistor have an amplitude that satisfies Eq. (4.3.3). In a later chapter, we will see other examples of clock generators and will discuss the design of the buffers.

Shift registers are used in the design of counters, frequency scalers, serial-in-parallel-out registers, and other circuits when a delay of the input signal is desired. Fig. 4.4.3 shows a simple frequency divider for moderate modulus N that consists of a chain of shift registers. When "mode select" is set to 1, the circuit is a divide-by-2N scaler. When it is set to 0, the circuit becomes a divide-by-(2N+1) scaler. Fig. 4.4.4 shows the computer simulation results of a divide-by-5 scaler that operates at a clock frequency of 2 GHz.

4.5 Cross Point Switch

As a second example, consider the cross point switch. It is a logic block with N inputs and M outputs. Data appearing at any of the N inputs can be directed (switched) to any or all of the M output terminals.

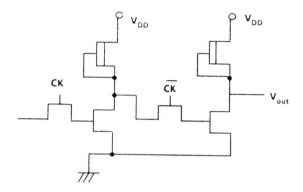

Fig. 4.4.1 Basic unit of a transmission-gate-based shift register.

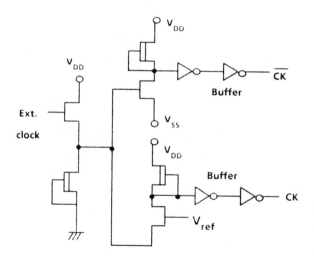

Fig. 4.4.2 A simple non-overlapping clock generator.

No two inputs may appear at the same output. Here, we will present a simple design in which each switch cell consists of three transmission gates. Later we will present more advanced designs.

Fig. 4.5.1 shows the block diagram of a 4x4 switch. Notice extra buffers are inserted after two cells so that each buffer drives only two cells. The control signals $C_{ij}, i, j = 1, ..., 4$ are separately generated according to external switching commands. For example, to direct data-1 to

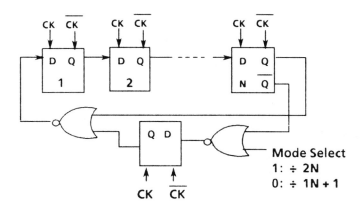

Mode Select
1: ÷ 2N
0: ÷ 1N + 1

Fig. 4.4.3 A simple divide-by-(2N)/(2N+1) frequency scaler.

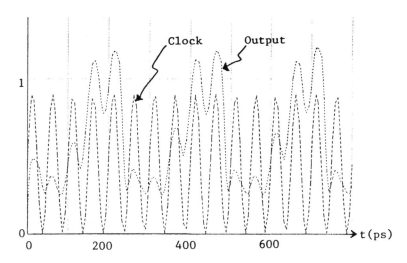

Fig. 4.4.4 Output waveform of a divide-by-5 scaler.

output-3, data-2 to outputs 1 and 4 and data-4 to output-2, C_{13}, C_{21}, C_{24}, and C_{42} must be high and all the other C_{ij} must be low. Note that the control signals along any column must be mutually exclusive.

The switch cell is shown in Fig. 4.5.2. All the transistors are of depletion mode. Suppose the data swings from 0.2 V and 0.9 V and the

threshold voltage of the transistors is -1.0 V. Then the control voltage
must be designed so that, according to Eqs. (4.3.3 and 4),

$$-0.1 < V_{GH} < 1.0 \qquad\qquad (4.5.1)$$

$$V_{GL} < -0.8 \qquad\qquad (4.5.2)$$

where we have assumed that $V_B = 0.8V$. A gate voltage that swings from
-1.0 to 0 V will do.

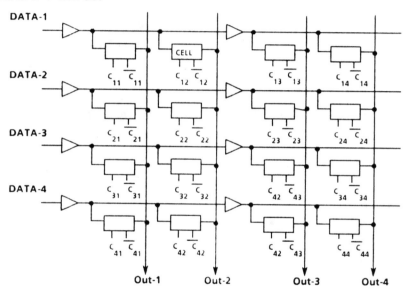

Fig. 4.5.1 A 4x4 cross point switch.

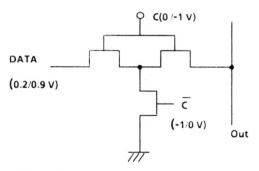

Fig. 4.5.2 A simple transmission-gate-based switch cell.

Chapter 5

Buffered ED Logic Circuits

Introduction

Logic circuits based on the enhancement-depletion inverter are simple in design and efficient in area and power. However, when the output is loaded with another stage, the maximum output voltage, V_{OH}, is limited to a value equal to the Schottky gate diode turn-on voltage, and the minimum output voltage, V_{OL}, is often greater than the threshold voltage of the enhancement transistor of the following stage. As a result, the noise margins are often comparable to the threshold voltage, so if the latter changes, the noise margins could become unacceptably small.

Noise margins can be improved by increasing V_{OH} or V_{IL}, by lowering V_{OL} or V_{IH}, or by doing a combination of these. (See Sec. 3.3.3.) In this chapter, we shall study circuits with improved noise margins. These circuits are composed of an ED inverter followed by a buffer. The buffer serves to lower the minimum output voltage V_{OL} or to increase V_{IL}. It also isolates the logic function of the circuit from the output and provides the necessary current to drive a large load either in the form of a capacitor or a number of fan-out transistors. The improvement comes at a cost of increased circuit complexity and delay.

We begin with an analysis of the source follower as a buffer.

5.1 Source Follower

A circuit in which the output follows the input except for a voltage drop
is the source follower shown in Fig. 5.1.1(a). The fan-out transistor
which the source follower drives is represented by a capacitor in parallel
with a diode. The capacitance C accounts for the total input capacitance
of the transistor and any parasitic and wiring capacitance, and it is
assumed to be linear.

DC Transfer Characteristics

Consider the DC performance of the circuit first. The depletion transistor
T_d provides negative feedback to T_e in that if the input V_{in} increases, the
current I_e and hence I_d increases, forcing V to increase. So the gate-to-
source voltage of T_e will remain about constant, thus keeping I_e constant. If V_{in} is kept below V_{DD}, the gate-to-drain diode will not conduct
and the gate-to-source voltage of T_e will be below V_B, the turn-on voltage of the gate-to-source diode. Thus there is no gate current drawn by
T_e, even when the input exceeds V_B, and the input voltage swing can be
much larger than that in an ED inverter.

If the input is sufficiently high, it is possible that the output voltage
exceeds the turn-on voltage of the gate-to-source diode of the following
stage and clamping will occur. Thus the maximum output voltage V_{OH}
is again limited to V_B. On the other hand, when the input is low, the output will drop to a value close to $V_{in} - V_T$. Thus it is possible to have a
minimum output voltage V_{OL} below the threshold voltage V_T.

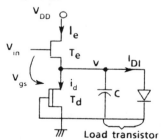

Fig. 5.1.1 A source follower

The transfer curve is shown in Fig. 5.1.2(b) and can be found
graphically. Fig. 5.1.2(a) shows the current I_e as a function of V for

different values of V_{in}. Plotted on the same graph is the current of the depletion transistor I_d. The intersections of these curves are the solutions of the circuit. We see from the transfer curve that at DC, the output follows the input up to some large value. At small values of input, the output is nearly zero.

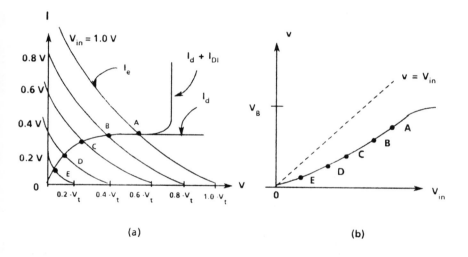

(a)   (b)

Fig. 5.1.2 (a) Graphical solution of source follower; (b) Transfer curve.

Pull-Up Delay

Next we turn to the pull-up and pull-down delays of the circuit. Let V_{in} be a pulse that rises from V_{IL} to V_{IH} in t_1 seconds. The circuit equation is:

$$(C+C_{gs})\frac{dV}{dt}=I_e-I_d-I_{DI}+I_{in} \qquad (5.1.1)$$

where

$$I_{in} = C_{gs}\frac{dV_{in}}{dt} = C_{gs}(V_{IH} - V_{IL})/t_1 \quad 0 < t < t_1$$

$$= 0, \quad elsewhere. \tag{5.1.2}$$

and C_{gs} is the gate-to-source capacitance of T_e. Ignoring I_{in} for the moment, we obtain the solution trajectory in the I-V plane shown in Fig. 5.1.3(a). (The curves for I_e for various V_{in} are shown in dotted lines.) At $t=0$, $V_{in}=V_{IL}$ and the solution begins at point A. At this point, $I_{DI} = 0$, $I_{in} = 0$, and V_{OL} is found from

$$I_e(V_{DD}, V_{IL}, V_{OL}) = I_d(V_{OL}, 0, 0) \tag{5.1.3}$$

by iteration. As V_{in} increases, I_e rises. At $t=t_1$, $V_{in}=V_{IH}$ and $V=V_1$ (point B). Thereafter, I_e follows the curve with $V_{in}=V_{IH}$. At some $V>V_1$, the diode begins to conduct and its current is added to I_d. At point C, $I_e=I_d+I_{DI}$ and the output reaches its steady state value V_B found from

$$I_e(V_{DD}, V_{IH}, V_B) = I_d(V_B, 0, 0) + I_{DI}(V_B) \tag{5.1.4}$$

by iteration again. The input and output waveforms are shown in Fig. 5.1.3(b) and (c).

Suppose there were no diode current. Then the solution trajectory would follow the I_e curve to a point C' in the steady state ($t = \infty$), and the output voltage will be V_{OH}. At some time t_2, the output voltage will be equal to V_B. We shall define the pull-up delay as t_2, namely as the time when $V = V_B$ under the condition that the gate current is zero. The differential equation from which the pull-up is computed is

$$(C + C_{gs})\frac{dV}{dt} = I_e - I_d + I_{in} \tag{5.1.5}$$

Integrating the equation over the two intervals $[0, t_1]$ and $[t_1, t_2]$ in the usual way, we get

$$t_{pu} = \frac{\left\{ 2(C + C_{gs})(V_B - V_{OL}) - 2C_{gs}(V_{IH} - V_{IL}) + t_1\left[I_e(V_{DD}, V_{IH}, V_B) - I_d(V_B, 0, 0) \right] \right\}}{I_e(V_{DD}, V_{IH}, V_B) - I_d(V_B, 0, 0) + I_e(V_{DD}, V_{IH}, V_1) - i_d(V_1, 0, 0)} \tag{5.1.6}$$

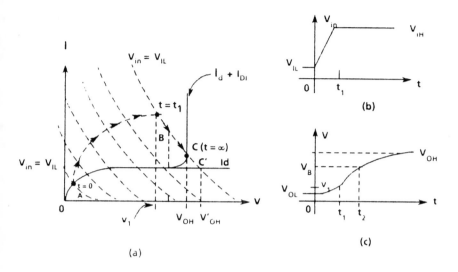

Fig. 5.1.3 (a) Solution trajectory on pull-up. (b) Input waveform.
(c) Output waveform.

where V_1 is given by

$$v_1 = V_{OL} + \frac{C_{gs}}{C + C_{gs}}(V_{IH} - V_{IL})$$

$$+ \frac{t_1}{2(C + C_{gs})}\left[I_e(V_{DD}, V_{IH}, V_1) - I_d(V_1, 0, 0)\right] \quad (5.1.7)$$

and must be solved by iteration.

From Eq. (5.1.6) we see that to reduce delay, the width ratio W_d/W_e should be made small. The capacitor is being charged through the enhancement transistor and any current taken by the depletion transistor reduces the amount of current available to charge the capacitor. The gate to source capacitance aids the charging current from the input source. The special case of a step input is obtained by setting t_1 to zero.

Pull-Down Delay

On pull-down, the capacitor discharge starts at point A in the I-V plane as shown in Fig. 5.1.4 where

$$I_e(V_{DD}, V_{IH}, V_{OH}) = I_d(V_{OH}, 0, 0) + I_{DI}(V_{OH}) \qquad (5.1.8)$$

from which we obtain V_{OH} by iteration.

As the input voltage V_{in} decreases, I_e decreases. If we assume that V_{IL} is so low and t_1 is so small that at some time t_0, the enhancement transistor is cut off, then the solution trajectory will follow the abscissa towards $V = 0$ at $t = \infty$, passing through V_1 at t_1, as shown in Fig. 5.1.4.

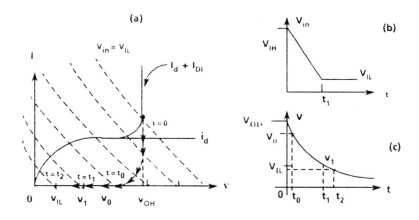

Fig. 5.1.4 (a) Solution trajectory on pull-down. (b) Input waveform. (c) Output waveform.

Let $V_0 \equiv V(t_0)$. Then by definition

$$V_{in}(t_0) - V_0 = V_{Te}$$

and from the input waveform, we have

$$V_{in}(t_0) = V_{IH} - (V_{IH} - V_{IL}) \frac{t_0}{t_1}$$

so

$$V_0 = V_{IH} - (V_{IH} - V_{IL})\frac{t_0}{t_1} - V_{Te} \qquad (5.1.9)$$

To find t_0, we integrate the circuit equation from $t=0$ to $t=t_0$ in the usual way and get

$$V_0 = V_{OH} - \frac{C_{gs}}{C + C_{gs}} \frac{V_{IH} - V_{IL}}{t_1} t_0 - \frac{t_0}{2(C + C_{gs})} I_d(v_0, 0, 0) \qquad (5.1.10)$$

Eqs. (5.1.9-10) are solve simultaneously to find t_0 and V_0.

Let $V_1 = V(t_1)$. Integrating over $[t_0, t_1]$ and noting that in this interval I_e and I_{DI} are zero, we get

$$V_1 = V_0 - \frac{t_1 - t_0}{2(C + C_{gs})} \left[I_d(V_1, 0, 0) + I_d(V_0, 0, 0) \right] \qquad (5.1.11)$$

from which we solve for V_1 by iteration.

Finally, the pull-down delay will be defined as the time when the output voltage reaches V_{IL} (arbitrary but reasonable). Integrating over $[t_1, t_2]$, we find the pull-down delay to be

$$t_{pd} = t_1 + \frac{2(C + C_{gs})(V_1 - V_{IL})}{I_d(V_1, 0, 0) + I_d(V_{IL}, 0, 0)} \qquad (5.1.12)$$

If the input is a step, $t_1=0$ and $V_1 = V_{OH}$. The pull-down delay reduces to

$$t_{pd} = \frac{2(C + C_{gs})(V_{OH} - V_{IL})}{I_d(V_{OH}, 0, 0) + I_d(V_{IL}, 0, 0)} \qquad (5.1.13)$$

It is seen that the pull-down delay can be reduced by making the depletion transistor large since the capacitor discharges through it. Once more, we see the conflicting requirements for optimal circuit design. In practice we choose the width ratio so that the pull-up and pull-down delays are about the same.

From the expressions for the delays, we can see the effects of fan-out. If the number of transistors driven by a source follower is FO, then to first order, the pull-up and pull-down delays will be proportional to FO.

OR Gate

The source follower can be used as an OR gate, as shown in Fig. 5.1.5. It should be noted that the logical 1 levels of the input and output are different.

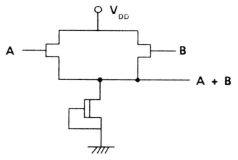

Fig. 5.1.5 A source-follower OR gate.

Simulation Results

Fig. 5.1.6(a) shows the computer simulation results of a source follower loaded by an inverter. The input is a pulse of amplitude 1.5 V, equal to V_{DD}. In this case, the input is sufficiently high so that the output is clamped at about 0.8 V. The width of the depletion transistor is varied from 7.5 to 20 microns to show how the pull-down delay is reduced as we make the width larger. However, the maximum output voltage is reduced as we make the depletion transistor larger, and this is undesirable. A good compromise is to choose a width ratio of about 1/3. The effect of fan-out is shown in Fig. 5.1.6(b). As expected, the delay increases as fan-out increases.

5.2 EDSF Logic

We will now utilize the two distinct properties of the source follower in the design of a buffered inverter that has better noise margins than the ED inverter. The two properties are that the input transistor of the source follower does not draw any gate current even when the input voltage is as high as V_{DD}, and that the minimum output voltage is usually lower than the threshold voltage of an enhancement transistor. The first such inverter is the EDSF inverter shown in Fig. 5.2.1. The name is

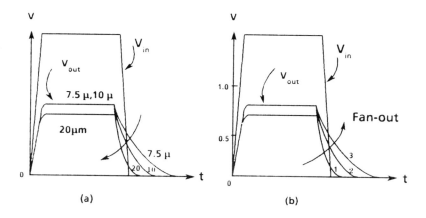

Fig. 5.1.6 (a) Pulse response, showing how the pull-down delay is reduced as we make the depletion transistor larger. (b) Effect of fan-out on delay.

derived from "ED inverter followed by a source follower."

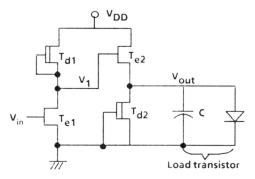

Fig. 5.2.1 An EDSF inverter.

DC Transfer Characteristics

Consider the situation when V_{in} is very low so that T_{e1} is essentially cut off and V_1 is close to V_{DD}. The output voltage V_{out} will be high and clamped to the barrier voltage of the input diode of the following stage. As V_{in} increases, V_1 drops and V follows. When $V_{in}=V_B$, V_1 reaches its minimum. When V_{in} exceeds V_B, the gate to drain diode of T_{e1} is turned on and V_1 begins to rise, and so does V_{out}. The transfer curve is shown in Fig. 5.2.2. Note that the maximum value of V_{out} is about 0.8 V and its minimum is about 0.1 V, significantly lower than that of an ED inverter. We expect therefore the low noise margin of the EDSF to be better. Fig. 5.2.3 shows the simulation results of the noise margins of an EDSF compared with those of an ED inverter. The width ratios of the EDSF are $W_{d1}/W_{e1}=7.5/15$ and $W_{d2}/W_{e2}=10/30$. For this example, an optimum design would call for a threshold voltage of the enhancement transistor of about 0.15 V.

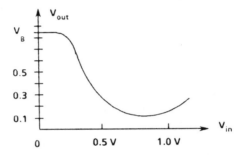

Fig. 5.2.2 Transfer curve of an EDSF inverter.

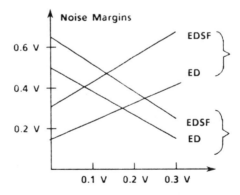

Fig. 5.2.3 Comparison of noise margins of an ED inverter with those of an EDSF inverter.

Delay

The delay can be computed by first computing the delay of the ED inverter and adding to it the delay of the source follower using the approximate output waveform of the ED inverter as the input to the source follower. The expressions of the delay are complex and will not be given. It is expected that the delay will be larger than the ED inverter. By simulation, we obtain the pulse response shown in Fig. 5.2.4. It is seen that the effect of threshold voltage change on delay is slight except that the voltage swing is reduced. The pull-down delay is increased as fan-out increases but the pull-down delay remain relatively unchanged. Note that the voltage swing of an EDSF inverter is from 0.17 V to 0.8 V, much larger than the swing in an ED inverter.

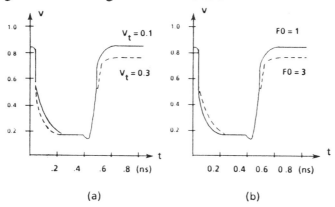

Fig. 5.2.4 Pulse response of EDSF inverter. (a) Effect of threshold voltage change. (b) Effect of fan-out.

EDSF NOR

Fig. 5.2.5(a) shows an EDSF NOR circuit. Note that the logic function is realized in the ED inverter and the source follower acts to isolate the logic from the output. The comments on ED NOR gates of Sec. 3.2 apply here, namely, the pull-up delay is increased as additional inputs are added to the NOR gate.

EDSF Buffer

In logic circuits there are occasions where a signal, e.g., a clock signal, must be applied to a large number of transistor gates, such as the gates of the transmission gates of a shift register. If the amplitude of the clock signal is limited to its allowable values, the gate diode of each

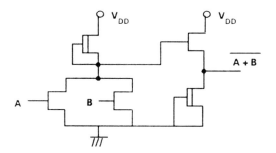

Fig. 5.2.5 An EDSF NOR gate.

transistor is not turned on and the load that presents itself to the source of
the clock is mainly capacitive. Its equivalent circuit is simply a capacitor
of large capacitance. Similarly, when a signal must be connected to
another chip, the signal line is usually terminated in a metallic "pad,"
which can be regarded as a large capacitance.

From the study of delay of circuits, we see that it can be expressed
as

$$delay = \frac{C\Delta V}{I} \qquad (5.2.1)$$

where C is some capacitance, usually the load plus wiring capacitance;
ΔV is the voltage swing and I is some current expression. If an inverter
drives another, then, to first order, C is proportional to the width of the
input transistor of the following stage. The current I, again to first order,
is proportional to the width of the transistor through which the load capacitor is charged or discharged. The voltage swing ΔV is fixed in most
cases, though it does depend on the widths of the transistors, but to
second order.

Eq. (5.2.1) suggests that to drive a large capacitive load, the transistor must be large. It is possible that the large width required is impractical. A solution to this problem is to use a chain of inverters with the
sizes of transistors progressively larger over the chain.

Consider the delay of an EDSF inverter. With reference to Fig. 5.2.1, let the width ratio W_{d1}/W_{e1} be fixed at, say, 1/2, and let W_{d2}/W_{e2} be fixed at 1/3. Then to first order, from the expressions of the pull-up and pull-down delays that we have derived so far, we can write

$$t_d = k_0 + k_1 \frac{W_{e2}}{W_{e1}} + k_2 \frac{W_{e3}}{W_{e2}} \qquad (5.2.2)$$

where W_{e3} is the width of the enhancement transistor of the ED inverter of the following stage. The constants k_0, k_1 and k_2 can be determined by simulation. The widths W_{e1} and W_{e2} can be chosen to optimize the delay performance of the circuit and a value of $W_{e1}/W_{e2}=1/2$ is a good choice. We will assume that this ratio is fixed and Eq. (5.2.2) is simplified to

$$t_d = m_0 + m_1 \frac{W_2}{W_1} \qquad (5.2.3)$$

where W_1 is W_{e1} of the driving stage and W_2 is W_{e1} of the driven stage, and m_0 and m_1 are constants. In a chain of N EDSF inverters, terminated in a capacitance C, we would have

$$t_d = m_0 + m_1 \frac{W_2}{W_1} + m_2 \frac{W_3}{W_2} + \cdots + m_N \frac{C}{W_N} \qquad (5.2.4)$$

For simplicity, we assume that $m_1 = m_2 = \cdots = m_N$. The delay is minimized if $W_2/W_1 = W_3/W_2 = \cdots = W_{N-1}/W_N$. Thus the widths $W_1, W_2, \cdots, W_N$ form a geometric series. Fig. 5.2.6 shows a design with W_2/W_1 equal to 1/2, and Fig. 5.2.7 shows the pulse response of such a buffer driving a capacitor of 0.1, 0.5 and 1pF.

5.3 EDSFD Logic

In the last section, we note that the transfer curve of an EDSF inverter increases as the input voltage increases beyond the turn-on voltage of the input gate diode, owing to the gate to drain current in the ED inverter. This rise in the transfer curve is undesirable for it could cause logic error. Secondly, when the input is low, the output will cause the input gate diode of the following stage to conduct. Both problems can be

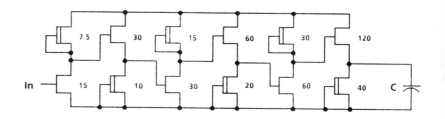

Fig. 5.2.6 An EDSF buffer.

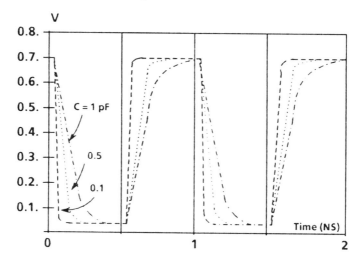

Fig. 5.2 7 Pulse response of an EDSF buffer driving a capacitor of 0.1,
0.5 and 1pF.

avoided if we insert a diode in series with the enhancement transistor of
the source follower as shown in Fig. 5.3.1.

When the input is low, the output will be one diode drop below V_2,
and V_2 itself is about 40% below V_1. Hence, it is possible to select
widths so that V_{out} is less than the turn-on voltage of the load diode.
When the input is high and V_1 is sufficiently low, the series diode is not
conducting and the output V_{out} drops to zero. In the transfer curve, we
expect $V_{OH} \leq V_B$ and $V_{OL} = 0$. Fig. 5.3.2 shows the simulation results as the

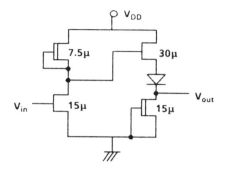

Fig. 5.3.1 An EDSFD inverter.

threshold voltage of both enhancement transistors varies.

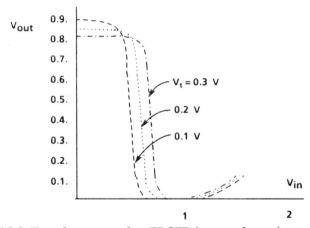

Fig. 5.3.2 Transfer curves of an EDSFD inverter for various values of
the threshold voltage of the enhancement transistors.

It is seen that the transfer curve has excellent symmetry and excep-
tionally good noise margins. This circuit is particularly suitable for use
in an inverter chain or shift register. The delay should be slightly better
than a basic EDSF inverter. On pull-up, the output voltage never
exceeds the turn-on voltage of the load diode so that all of the current
from the source follower goes to charge the load capacitor and none to
the diode. On pull-down, all of the capacitor current goes to the

depletion transistor T_{d2} and the discharge current is not reduced by the current from the enhancement transistor T_{e2}. Fig. 5.3.3 shows the simulation results of the pulse response of an EDSFD inverter with fan-out equal to one, two and three. It is seen that the pull-up delay is almost independent of fan-out and that the minimum output voltage reaches zero, as expected. Thus the EDSFD inverter has a large voltage swing without a corresponding increase in delay.

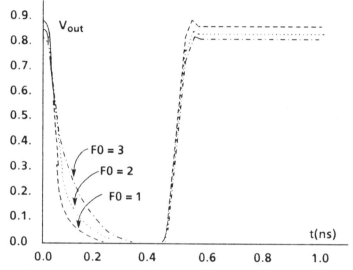

Fig. 5.3.3 Pulse response of EDSFD inverter for fan-out equal to 1,2, and 3.

5.4 EDSB Logic

In an EDSF inverter, the depletion transistor of the source follower acts as a fixed nonlinear resistor whose "resistance" is a function of its drain voltage only and is not controlled by any external signal. The performance of the inverter would be improved if we could make the nonlinear resistance smaller as we discharge the load capacitor through it so as to reduce the pull-down delay, and to make it larger as we charge the capacitor through the enhancement transistor so as to reduce the current taken away from the charging current. Two circuits that accomplish this goal are the ED super-buffers shown in Fig. 5.4.1(a) and (b), one being "inverting" and the other not. The input stage is an ED inverter and the output stage consists of two enhancement transistors controlled by the input and its complement so that as one input goes high the other goes low, and the corresponding transistor "resistances" are increased or decreased, as the load capacitor is being charged or discharged.

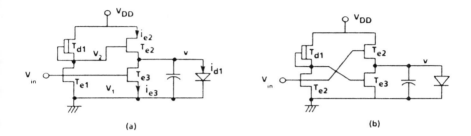

Fig. 5.4.1 (a) Inverting ED super-buffer. (b) Non-inverting ED super-buffer.

We shall concentrate on the inverting super-buffer. When the input is low, V_2 will be close to V_{DD}. At the same time, transistor T_{e3} will be less conductive. These two consequences reinforce each other so that the output voltage V becomes high and is clamped to V_B over a range of input voltage values. In the transfer curve, V_{IL} will be larger than that of an ED inverter and the low noise margin is improved. When the input is high, V_2 is low, T_{e2} becomes less conductive while T_{e3} becomes highly conductive, and the output voltage V drops to a very low value, again improving the low noise margin.

DC Characteristics

The transfer curve can be obtained graphically. Because of the negative feedback provided by T_{e3} to T_{e2}, the gate current of T_{e2} will be negligible and the ED inverter can be treated separately. The transfer curve of the ED inverter is shown in Fig. 5.4.2 where we note that V_{OH} attains the value of V_{DD} We denote the values of V_2 at various values of V_{in} by $V_2(V_{in})$ as points A, B, ..., E. For the output stage, we have plotted the current I_{e2} and I_{e3} in the I-V plane for various values of V_2 and V_{in}, respectively, as shown in Fig. 5.4.3.

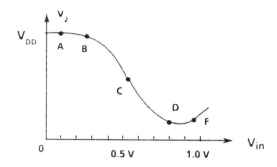

Fig. 5.4.2 The transfer curve of the ED inverter of the EDSB.

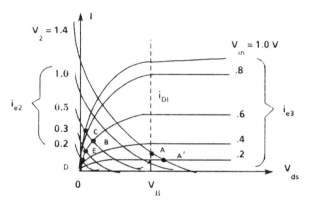

Fig. 5.4.3 I-V characteristics of T_{e2} and T_{e3} of the EDSB.

The solution points are indicated as A,B,...,E. If the buffer drives another stage, then point A' is not possible since V would be clamped to a value V_B at point A. It is seen that $V_{OH} \approx V_B$ and $V_{OL} \approx 0$. However, since the input drives both T_{e1} and T_{e3}, when V_{in} is high enough to cause the gate-to-drain diode in each of the transistors to conduct, V_2 and V rise together rapidly. For this reason, the input voltage must be limited to a value not appreciably larger than V_B. Fig. 5.4.4 shows the transfer curve and simulation results for various values of the threshold voltage of the enhancement transistor. Note that the low noise margin is very good but the high noise margin is not because of the rapid rise of the output as the input increases.

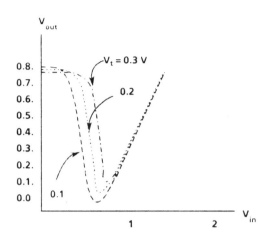

Fig. 5.4.4 Transfer characteristics of an EDSB.

Delay

The driving capability of the super-buffer can be appreciated by studying the pull-up and pull-down delay characteristics and compare them with those of an EDSF inverter. Fig. 5.4.5 shows the results of simulation of one stage of a super-buffer terminated in a capacitor, compared with those of one stage of an EDSF inverter identically terminated. The width ratios of the ED super-buffer are: $W_{d1}/W_{e1}=7.5/15$ and $W_{e2}/W_{e3}=30/30$. Those of the EDSF inverter are $W_{d1}/W_{e1}=7.5/15$ and $W_{d2}/W_{e2}=10/30$. The super-buffer is superior. However, we should notice that in a super-buffer, the input drives two enhancement transistors with a capacitance corresponding to a transistor width of 45 units. This is why the width of T_{e2} is as large as that of T_{e3}.

The fan-out capability can be discerned from the pulse response shown in Fig. 5.4.6 when the buffer is loaded with one, two or three identical buffers. The pull-up delay is almost independent of fan-out.

NOR Gate

An EDSB NOR gate with three inputs is shown in Fig. 5.4.7. Since each input must drive two transistors, the area requirement is considerably larger than that of an EDSF NOR. In addition, the two transistors

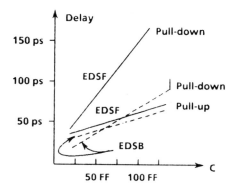

Fig. 5.4.5 Comparison of pull-up and pull-down delays of EDSB and EDSF buffers driving a capacitor.

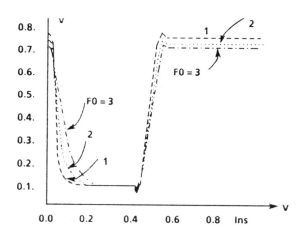

Fig. 5.4.6 Pulse response of EDSB for different values of fan-out.

present a capacitive load to the driving stage about three times as large as that presented by an EDSF. For these reasons, the EDSB inverters are ordinarily used as buffers and not as logic elements.

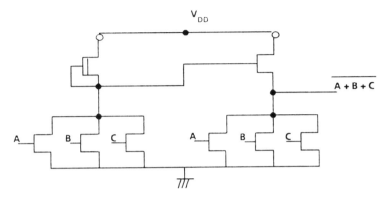

Fig. 5.4.7 An EDSB NOR gate.

5.5 SFED Logic

In all of the inverter circuits we have studied so far, the input stage is an ED inverter. When a buffered inverter with an ED inverter as the input stage is loaded by another, the output voltage is limited to a value equal to the Schottky barrier voltage because of gate conduction. It would be desirable if this limitation is removed. We recall that a source follower has the property that its input transistor does not draw any gate current even when the input voltage is as high as V_{DD}, and its minimum output voltage can be close to zero. This suggests that we should use a source follower as the input stage and an ED inverter as an output stage to make up a buffered inverter. Such an inverter is shown in Fig. 5.5.1 and will be called a SFED inverter.

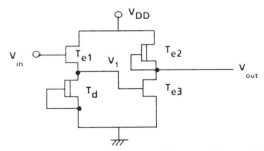

Fig. 5.5.1 An SFED inverter (a source follower followed by an ED).

DC Characteristics

When the input is low, V_1 will be quite low so that transistor T_{e3} will be cut off and the output voltage will reach V_{DD}. When the input is high but below V_{DD}, V_1 will be a few tenths of a volt below the input or at most equal to V_B. There is therefore no gate current drawn by T_{e3} between the gate and drain. The output will not rise as in an ED inverter. We expect this inverter to have a large voltage swing and good noise margins.

Fig. 5.5.2 shows the simulation results of the transfer characteristics of a SFED inverter for different values of the threshold voltage of the enhancement transistor V_T. Note the absence of any rise of the output as the input increases to V_{DD} and the symmetry of the transfer curve. At $V_T=0.1V$, the low and high noise margins are almost equal and have a value of about 0.23 V. The maximum output voltage is V_{DD}, as expected.

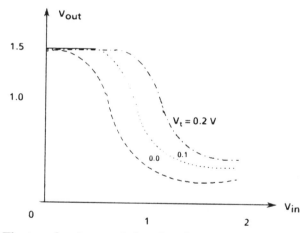

Fig. 5.5.2 The transfer characteristics of an SFED inverter for different values of the threshold voltage of the enhancement transistor.

Delay

According to linear circuit theory, a source follower has the property that its input impedance is very high, in that any impedance connected across the gate and source is amplified by a factor of $(1 + g_m R_S)$, where g_m is the small signal transconductance and R_S is the small signal equivalent resistance of the depletion transistor. Though linear circuit theory does not apply to large signal digital circuits, it does suggest that in the nonlinear case the input capacitance of a source follower nonlinear case should be smaller than that of an ED inverter. It follows that the pull-up and pull-down delays of an SFED inverter should be fairly independent of fan-out. This is confirmed by simulation as shown in Fig. 5.5.3. Note the large voltage swing that is possible with this circuit.

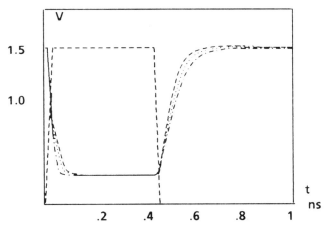

Fig. 5.5.3 Pulse response of an SFED inverter for various values of fan-out.

5.6 Split-Phase Super Buffer

We recall that in a super-buffer, the output stage is driven by two signals that are complementary, or electrically 180° out of phase with respect to each other. The performance of the circuit is improved because as the pull-up transistor in the output stage is conducting heavily, the pull-down transistor is near cutoff, and vice versa. The result is that the pull-up and pull-down delays are reduced. Its fan-out capability is much better as well. However, the high noise margin of the super buffer is poor because of the gate-to-drain conduction of the input transistor. In addition, when the output stage is loaded by another such inverter, the output voltage is limited by the Schottky voltage and the voltage swing is at most equal to V_B. These problems can be overcome by using a phase splitter as the input stage to generate the two complementary signals and retaining the push-pull arrangement of the output stage, as shown in Fig. 5.6.1 [5.1,2].

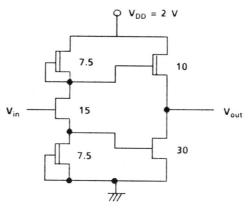

Fig. 5.6.1 A split-phase super buffer inverter and a NOR gate.

The voltages V_1 and V_2 are electrically complementary as desired. They drive the output stage, which serves as a buffer. Because the input transistor is connected to a load at both its drain and source, the input voltage can rise above V_B without drawing any gate current, so we expect the voltage swing to be large. Fig. 5.6.2 shows the transfer curve of the inverter and it is seen the voltage swing is about $1.2V$, significantly larger than that of an EDSB. The noise margins are excellent, being $300mV$ for both the high and low margins, and are much larger than those of an ED inverter..

The pulse responses are shown in Fig. 5.6.3 for three values of fan-out. We see that the pull-down delay is almost independent of fan-out and that the voltage swing is as large as that of the input.

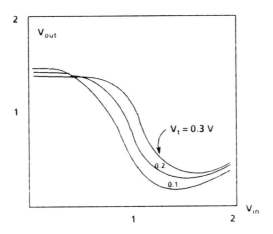

Fig. 5.6.2 Transfer curve of a split-phase super buffer.

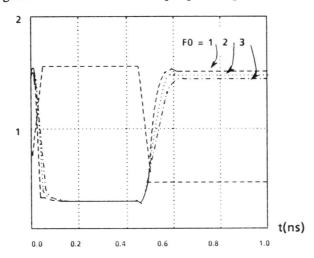

Fig. 5.6.3 Pulse response of the split-phase super-buffer.

5.7 Ring Oscillator

In the discussion of pull-up and pull-down delays, we defined delay as the time taken for the output to go from one voltage level to another. For the purpose of deriving analytic formulas, we chose the voltage levels to be V_{OH} and V_{OL}. However, when one is confronted with an actual waveform, it is often difficult to decide what the proper voltage levels might be. The fact is that there is no agreement among practitioners as to how delay should be defined or measured. Many definitions are used in practice depending more on convenience than on universal applicability.

There is, however, a practical way to compare circuit performance with respect to speed or delay, and it is the use of a ring oscillator. The oscillator consists of a ring of N identical inverters as shown in Fig. 5.7.1. With N odd and provided that each inverter has sufficient gain, the circuit will sustain oscillation and produce a periodic waveform of some period T, which will be shown later to be related to the inverter delay. In practice, the harmonics of the waveform are measured with a spectrum analyzer and from the fundamental frequency, the period, and hence the delay, is determined.

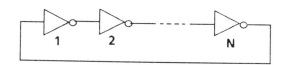

Fig. 5.7.1 A ring oscillator consisting of N stages of inverters.

Analysis

First we should establish that the number of inverters in a ring must be odd. If N is even, then the circuit will be latched into a stable state where the inverter output will be alternately high or low and there will be no oscillation. With N odd, the input to the first inverter, which is the output of the last, changes after some delay. The delay will adjust itself so that a steady oscillation is sustained.

To analyze the circuit, we need to look for a periodic solution to a system of nonlinear differential equations that describes the circuit. There is no simple, analytic way to do this. Instead, we shall use the technique of small signal analysis to determine the conditions that must be satisfied by the device parameters in order to support a steady oscillation in the circuit.

From the pulse response of all of the inverters that we have presented so far, we see that the response is very similar to that of a simple RC circuit. Thus, to first order, each inverter may be represented by a small signal equivalent circuit shown in Fig. 5.7.2. When N such circuits are connected into a ring, the following circuit equation must be satisfied.

$$v_1 = - \frac{g_m^N}{(g + j\omega C)^N} v_1 \qquad (5.7.1)$$

or

$$\frac{g}{g_m} + \frac{j\omega_k C}{g_m} = \exp(-j(2k-1)\pi/N)$$

Separating into real and imaginary parts, we have

$$\frac{g}{g_m} = \cos \frac{(2k-1)\pi}{N} \qquad (5.7.2)$$

$$\frac{\omega_k C}{g_m} = \sin \frac{(2k-1)\pi}{N} \qquad (5.7.3)$$

where $k = 1, 2, ..., (N-1)/2$. Each value of k corresponds to a frequency of oscillation, provided that there exists a value of g and g_m to satisfy Eq. (5.7.2). The output waveform will not be sinusoidal but contains harmonics. Let the fundamental frequency be ω_1 and the fundamental period be T_1. Then from Eqs. (5.7.2) and (5.6.3), we have

$$RC = \frac{1}{\omega_1} \tan \frac{\pi}{N} \qquad (5.7.4)$$

where $R = 1/g$. If N is large, then

$$RC \approx \frac{\pi}{\omega_1 N} = \frac{T_1}{2N} \qquad (5.7.5)$$

Thus the time constant of each stage is related to the period of oscillation in a simple way. If we loosely define the delay of each gate as two times the time constant (pull-up plus pull-down) of the linearized circuit, then we have

$$Gate-delay = \frac{T_1}{N} \qquad (5.7.6)$$

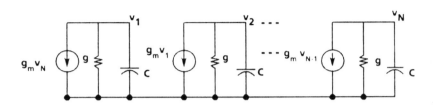

Fig. 5.7.2 Linearized, small-signal analysis of a ring oscillator.

Eq. (5.7.6) provides a basis to compare circuit performance. A ring oscillator is constructed of N identical inverters and its fundamental period is determined. The period divided by N gives an estimate of the inverter delay per stage. On this basis, we can also compare circuits by simulation. Fig. 5.7.3 shows the delay of various inverters plotted as a function of power dissipation per stage obtained from simulation of a 9-stage ring oscillator. Each stage is loaded with a capacitance of 50FF to model the interconnect capacitance. It is seen that the ED super-buffer is the fastest and the ED inverter consumes the least amount of power, as expected.

We can also use the ring oscillator to determine the fan-out and fan-in capability of an inverter. Each inverter is loaded with K identical inverters and the delay per stage is found for various values of K to determine the delay as a function of fan-out. Similarly, N NOR gates each with K fan-in's can be connected into a ring to determine the effect of fan-in on delay.

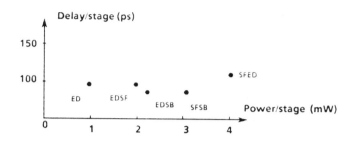

Fig. 5.7.3   Power-delay relation of various types of inverters.

5.8 Summary

In this chapter, we have presented a number of buffered inverters that have significantly better noise margins and fan-out capability than the basic ED inverter. Each buffered inverter consists of two stages, one to realize the inverter or NOR function and the other to provide the driving capability. The added complexity improves the noise margins at the expense of increased power consumption. As the simulation results of Fig. 5.7.3 show, the delay performance is only slightly inferior to that of an ED inverter. Buffered logic is therefore a practical and advantageous alternative to the ED logic if power is not a major design consideration. The following table summarizes the key features of the inverters.

Inverter	NML	NMH	V_{OL}	V_{OH}	Delay	Power
ED	Low	Low	$>V_T$	V_B	Low	Low
EDSF	Low	High	$\approx V_T$	V_B	Mod	Mod
EDSFD	High	High	0	V_B	Mod	Mod
EDSB	High	Low	$>V_T$	V_B	Low	Mod
SFED	High	High	$>V_T$	V_{DD}	High	High
SPSB	High	High	$>V_T$	$>V_B$	Mod	Mod

References

[1]   W.B. Leung, Y.K. Lo, Y.T. Oh, W.A. Oswald, E.K. Poon, C.E. Reid, L.E. Ackner, and T.C. Poon, "2K Gate circuits with 125ps Gate Delay Using GaAs HEMT Technology," IEEE GaAs Integrated Circuits Symposium Technical Digest, 1989, pp. 57-60.

[2]   W.B. Leung, K.W. Teng, A.I. Faris, A.C. Hu, L.E. ACkner, C.E. Reid, and T.C. Poon, "3.5K Gate 32-Bit ALU Using GaAs HFET Technology." Private communication.

Chapter 6

Source-Coupled Logic Circuits

Introduction

We recall from Chapter 2 that the threshold voltage of a GaAs MESFET is the difference of the Schottky barrier voltage and the pinch-off voltage, the latter being proportional to the square of the thickness of the active layer, so that the threshold voltage is very sensitive to geometric variations. At the present time, the processing technology is such that it is difficult to limit the variation of the threshold voltage to within $\pm 20\,mV$ while its nominal value is $0.1\,V$ for an enhancement transistor and approximately $-1.0\,V$ for a depletion transistor. From Chapters 3 and 5, we saw how the transfer curve and noise margins are affected by variations of the threshold voltage. In all cases, the circuit threshold shifts by an amount approximately equal to the change in the threshold voltage.

A circuit which has an inherent property that its circuit threshold voltage is independent of the threshold voltage of the transistors is the source-coupled differential pair. Moreover, it has excellent noise margins and the circuit can be so arranged that the voltage swing is significantly larger than the Schottky barrier voltage.

In this chapter, we shall present the GaAs MESFET source-coupled logic family, describe a graphical method of analysis and study its performance.

6.1 Source-coupled Differential Pair

A MESFET differential pair is shown in Fig. 6.1.1. We shall first obtain its transfer curves V_{o1} vs V_1 and V_{o2} vs V_1, for two cases: (1) $V_2 = V_{ref} = constant$, and (2) $V_2 = V_{ref}-V_1$. In the second case, V_1 and V_2 are electrically and logically complementary.

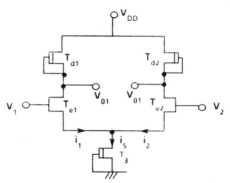

Fig. 6.1.1 A source-coupled differential pair.

Let the relevant node voltages and transistor currents be denoted as given in Fig. 6.1.1. To understand the operation of the circuit, a graphical solution, instead of computer simulation, is best. We first note that if we refer all the node voltages to V_S, then the circuit can be separated into two disconnected ED inverters. The solutions of the inverters can be separately obtained. They are coupled to each other through the requirement that $I_1 + I_2 = I_S$ and I_S is determined by V_S.

Consider case (1) first. For a given input voltage V_1, let the solution of the circuit be denoted as $I_1^*, I_2^*, V_S^*, V_{DS1}^*, V_{DS2}^*, V_{o1}^*$, and V_{o2}^*. Fig. 6.1.2 shows how the solution is obtained. In (a), we have plotted I_1 versus V_{DS1} with $V_{GS1} = V_1 - V_S$ as the usual gate-to-source voltage label on each curve. Similarly for I_2 in (b). The I-V curve for T_{d1} starts at $V_{DD}-V_S^*$ and its intersection with the I-V curve of T_{e1} for $V_{GS}=V_1-V_S^*$ defines the solution point for the ED inverter on the left. From this point, we obtain the drain current I_1^* and the output voltage $V_{o1}^* = V_{DS1}^* + V_S^*$. Similarly, we obtain I_2^* and V_{o2}^* from (b). The drain currents I_1^* and I_2^* must be such that their sum $I_1^* + I_2^*$ equals in fact I_S^*, the current of T_{d3} at a drain voltage equal to V_S^*. See Fig. 6.1.2(c). From this description of the solution, we obtain an iterative process to find V_{o1} and V_{o2} for each value of input V_1, as follows.

(1) For a given V_1, assume a value for V_S.

(2) With this value of V_S, draw the I-V curve for I_{d1} in the I-V plane of T_{e1}. Draw the I-V curve of I_{d2} in the I-V plane of T_{e2}.

(3) Identify the I_1 curve corresponding to $V_{GS} = V_1 - V_S$. The intersection of this curve with the I_{d1} curve gives I_1 and V_{DS1}. Similarly, we obtain I_2 and V_{DS2} from the I-V plane of T_{e2}.

(4) Let $I_3 = I_1 + I_2$. From the I-V curve of T_3, we get V_3 corresponding to I_3. If $V_3 = V_S$, we have found a solution. The output voltages V_{o1} and V_{o2} are found from $V_{DS1} + V_S$ and $V_{DS2} + V_S$, respectively. If $V_3 \neq V_S$, change the assumed value V_S and repeat until a solution is found.

By following this procedure, we get the transfer curves of a typical differential pair given in Fig. 6.1.3 for both cases.

6.2 Observations

The following are noteworthy.

(1) For $V_1 \gg V_2$, T_{e1} is conducting heavily and V_S tends to rise. Since V_2 is held constant (case 1), V_{GS} of T_{e2} is reduced and I_2 is decreased, thus tending to keep V_S constant at a value so that T_{e2} is never cut off. So the maximum value of V_{o2} is less than V_{DD}.

(2) For $V_1 \ll V_2$, T_{e1} is cut off. V_{o1} goes to V_{DD} since T_{e2} does not draw much gate current because of the negative feedback provided by T_3. V_S is lowered since the only current through T_3 is from T_{e2}.

(3) At the cross-over, $V_1 = V_2$, $V_{o1} = V_{o2}$, and $I_1 = I_2$.

(4) If V_2 is held constant, the two transfer curves are not symmetrical, namely, V_{OH} and V_{OL} of the two curves are not the same.

(5) In order to produce a symmetrical output, the input voltages must be electrically complementary, namely $V_1 + V_2 = V_{ref}$, where V_{ref} is a sufficiently large voltage. Thus, if we wish to design a logic circuit made of source-coupled gates, every variable and its complement must be available (dual rail logic).

(6) In the graphical analysis, we have neglected gate current. Simulation results show that the effects of gate current are similar to those of

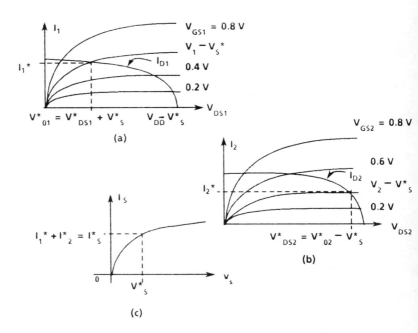

Fig. 6.1.2 (a) Solution point I_1^*, V_{DS1}^* of the left inverter of the source-coupled pair; (b) solution point I_2^*, V_{DS2}^* of the right inverter; (c) solution point I_S^*, V_S^* of transistor T_3.

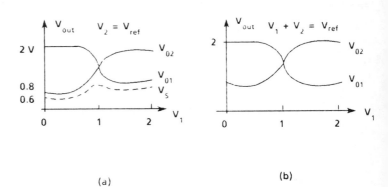

Fig. 6.1.3 Transfer curves of a source-coupled differential pair.
(a) $V_2 = V_{ref}$; (b) $V_2 = V_{ref} - V_1$.

an ED inverter. The reason is that V_S is fairly constant outside the cross-over region. As V_1 increases beyond the cross-over region, the drain voltage V_{o1} rises due to the large gate-to-drain current, as in an ED inverter.

(7) The minimum output voltage V_{OL} is relatively high. A source follower can be used at the output as a buffer to shift V_{OL} to a lower value, as shown in Fig. 6.2.1.

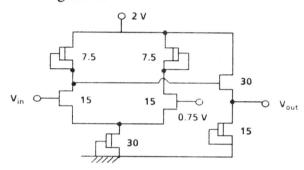

Fig. 6.2.1 A source-coupled inverter buffered by source-followers.

6.3 Noise Margins and Delays

Approximate analytic formulas of the noise margins and delays are possible, but the derivations are rather lengthy and will be omitted.

Fig. 6.3.1 shows the transfer curves obtained from computer simulation for a source-coupled pair buffered by a source follower. It is seen that the noise margins are excellent. In Fig. 6.3.2, we compare the noise margins of source-coupled inverter with those of an ED inverter as functions of the threshold voltage of the enhancement transistor. The delay performance as a function of fan-out (with the source follower as a buffer) is shown in Fig. 6.3.3.

6.4 Threshold Variation

Let us consider the effects of variations of the threshold voltage of the enhancement transistor on the circuit threshold voltage. We assume that the currents I_1 and I_2 can be written as

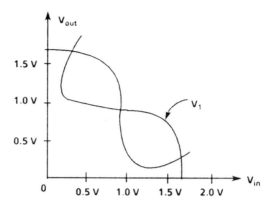

Fig. 6.3.1 Transfer curves of a source-coupled inverter.

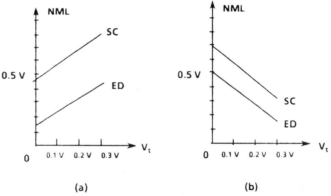

(a)   (b)

Fig. 6.3.2 Comparison of noise margins of a source-coupled inverter and
an ED inverter. (a) Low noise margin. (b) High noise margin.

$$I_1 = \beta_1 g (V_{GS1} - V_{t1}) f (V_{DS1}) \qquad (6.4.1)$$

$$I_2 = \beta_2 g (V_{GS2} - V_{t2}) f (V_{DS2}) \qquad (6.4.2)$$

where $g(.)$ and $f(.)$ are some single-valued functions of their argument.
V_{GS1} is the gate-to-source voltage of T_{e1} and V_{t1} its threshold voltage,
and V_{DS1} its drain-to-source voltage. (See Fig. 6.1.1.) Similarly for
those of T_{e2}. We define the circuit threshold voltage as the cross-over
voltage. At the cross-over, $V_{o1} = V_{o2}$, so $V_{DS1} = V_{DS2}$ and $I_1 = I_2$.
Therefore, we have, assuming $\beta_1 = \beta_2$:

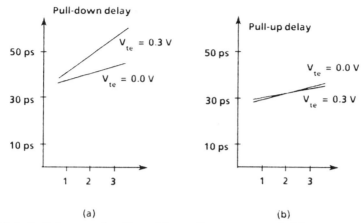

Fig. 6.3.3 Pull-down and pull-up delays of a source-coupled inverter
as a function of fan-out.

$$V_{GS1}-V_{t1} = V_{GS2}-V_{t2} \qquad (6.4.3)$$

or

$$V_1-V_S-V_{t1} = V_2-V_S-V_{t2} \qquad (6.4.4)$$

In the case $V_2 = V_{ref}$, we have that the circuit threshold V_{in}^* is given by

$$V_{in}^* = V_{ref} + V_{t1}-V_{t2} \qquad (6.4.5)$$

and if $V_1 + V_2 = V_{ref}$, then

$$2V_{in}^* = V_{ref} + V_{t1}-V_{t2} \qquad (6.4.6)$$

In either case, the circuit threshold is independent of the device threshold voltage if $V_{t1}=V_{t2}$ and if the change in V_{t1} tracks the change in V_{t2}, a condition to be expected since the two transistors are likely to be adjacent to each other on a chip. Fig. 6.4.1 shows the simulation results in which the transfer curves of a differential pair are plotted for two different values of the threshold voltage of the enhancement transistors. As evident from the figures, the cross-over voltage, or circuit threshold, does not change as the threshold voltage changes from 0.1 V to 0.3 V, though the logical low level of the non-inverted output is raised and the logical high level lowered when the threshold voltage is increased.

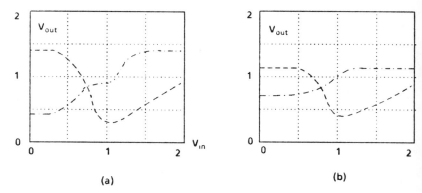

Fig. 6.4.1 The transfer curves of a source-coupled inverter.
(a) Inverted and non-inverted output, $V_{te}=0.1V$.
(b) Inverted and non-inverted output, $V_{te}=0.3V$.

6.5 Source-Coupled Logic

We saw in Sec. 6.2 that in order to obtain a pair of complementary outputs, the two inputs must also be complementary. This has advantages and disadvantages. A logic variable and its complement must be available throughout the circuit and the logic gates tend to be more complicated than ED logic. However, "series gating", commonly used in ECL, can also be used. Fig. 6.5.1 shows a 2-input NOR/OR gate, Fig. 6.5.2 an XOR/XNOR gate, and Fig. 6.5.3 a D-flip-flop.

In order to turn on the series-connected transistors, the logical 1 and 0 levels of the inputs will be different and must be maintained or adjusted by level shifters, such as source followers. Additional circuits will therefore be required and they take up chip area. As to switching speed, it appears that a fairly complicated function can be realized with only one "gate delay," in contrast with a chain of logic stages, which imposes several units of delays on the signal as it traverses through the chain. However, the gain in speed of the series gating circuits could be deceptive since the output capacitor must discharge through a number of series-connected transistors, so the pull-down delay could be large.

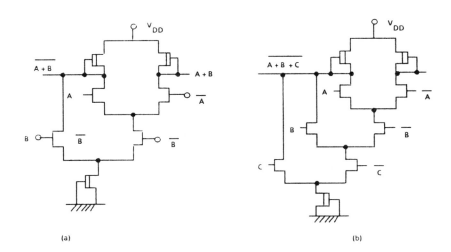

Fig. 6.5.1 (a) A 2-input source-coupled NOR gate with dual rail logic.
(b) A 3-input NOR.

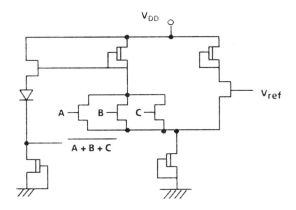

Fig. 6.5.2 A 3-input NOR with single-rail logic.

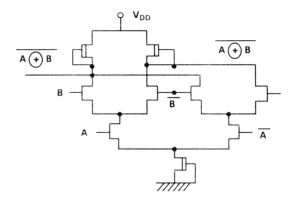

Fig. 6.5.3 A source-coupled XOR and XNOR.

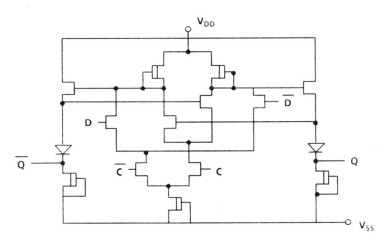

Fig. 6.5.4 A source-coupled D-flip-flop.

6.6 Cascode Differential Pair

We conclude this chapter by presenting a differential pair consisting of a cascode amplifier in each leg, as shown in Fig. 6.6.1. The circuit has large voltage swing, is insensitive to threshold variation, and has excellent noise margins and relatively large incremental gain. It is used as a comparator in an A/D converter [1].

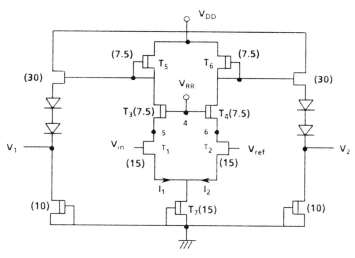

Fig. 6.6.1 A cascode differential pair.

The transfer curves are shown in Fig. 6.6.2 and it is seen that the inverted and non-inverted outputs are symmetrical ($V_t=0.1V$). The logical 1 levels of the inverted and non-inverted output are the same, so are the logical 0 levels, even though the input is only single-rail. The excellent noise margins and large voltage swing should be noticed. Notice also the absence of the rise of the output due to gate-to-drain conduction, as in an ED inverter when the input increases.

When the threshold voltage of the enhancement transistors is changed to $0.3V$, the symmetry of the transfer curves is lost but the circuit threshold is unchanged, as shown in Fig. 6.6.3. The inverted output still retains large noise margins and large swing.

The operation of the circuit can be explained as follows. When the input is low, T_1 is cut off. V_2 goes to V_{DD} and V_5 goes to a value such

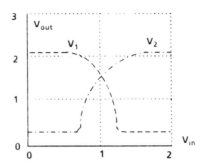

Fig. 6.6.2 Transfer curves of the cascode differential pair. Inverted (a) and non-inverted (b) output for $V_i=0.1V$.

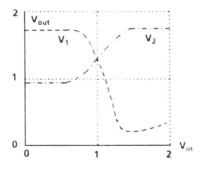

Fig. 6.6.3 Transfer curve of cascode differential pair for $V_i=0.3V$.

that the gate-to-source voltage of T_3, $V_{4,5}$ is just below the threshold voltage of T_3. At the same time, since T_1 is cut off, the current in T_7 comes only from T_2, so V_7 will be low. The gate-to-source voltage of T_2 is relatively large and T_2 conducts heavily. V_3 and V_6 will be low. Both remain low at a constant value independent of the input V_1 until V_1 increases to a value where T_1 begins to conduct. At this point, V_5 begins to drop and the gate-to-source voltage of T_3 increases and conducts heavily. The increased current causes V_2 to drop rapidly until at some value of V_1, and hence V_5, the current of T_3 saturates with respect to its gate-to-source voltage $V_{4,5}$, namely the current is constant independent of $V_{4,5}$, and V_2 also becomes constant. As V_1 increases further, the current I_1 increases and V_7 rises. When V_7 is sufficiently large, T_2 will be cut off, and V_3 rises to V_{DD} and V_6 to a value such that T_4 is cut off.

The pulse response is shown in Fig. 6.6.4 and it shows that large voltage swing is possible. Simulation results of a 9-stage ring oscillator show that the delay per stage is about 118ps, which is not appreciably larger than that of any of the buffered ED inverters.

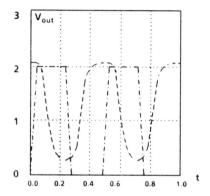

Fig. 6.6.4 Pulse response of cascode differential pair with the outputs
terminated in a 100FF capacitor.

6.7 Summary

The source-coupled differential pair is an attractive alternative to the ED or buffered inverters. Its insensitivity to device threshold change and its large noise margins and large logic swing are desirable features. However, it consumes relatively large amount of power since the circuit as a logic gate is never cut off and it is more complex and hence takes up more area than the simpler ED or buffered ED logic.

Reference

[1] T. Nguyen, F. Thomas, S. Ruggeri, M. Le Paih, J.M. Uro, F. Debrie, T. Dean and M. Gloanec, "A 4-bit full Nyquist 1 Gsample/s Monolithic GaAs ADC with on-chip S/H and error correction," IEEE GaAs Integrated Circuits Symposium Technical Digest, 1988, pp. 195-198.

Chapter 7

Subsystems Design

7.1 Introduction

In the preceding four chapters, we presented the analysis and design of GaAs logic circuits of different families: enhancement-depletion (ED) logic, transmission-gate logic, buffered ED logic, and source-coupled logic. In this concluding chapter, we shall present examples of systems design based on the GaAs technology. From the previous studies, four characteristics of GaAs circuits can be identified that are important in system design.

First, the switching speed of a GaAs gate is about an order higher than the speed of a Si MOSFET gate of comparable size. Typical values of pull-up and pull-down delays range from tens of picoseconds in a simple ED inverter to 120 ps in a complicated gate such as the source-coupled cascode differential pair. With gate delay of the order of 100 ps, GaAs circuits are most suitable for system applications that operate in the Gigabits per second range.

Second, unlike MOS circuits in which a complex logic function can be realized in one single stage, a stage of GaAs logic can only be either an inverter or NOR gate. All system logic functions must be constructed using only these two logic elements, plus possibly the pass transistor.

Third, at the present time, the processing technology is such that it is difficult to produce transistors within a chip with uniform quality. The variations of the important electrical parameters such as the threshold voltage, transconductrance, and drain conductance are relatively large compared to Si MOS devices. In deciding which of the many logic families to use in a system design, we must take into consideration the trade-offs of noise margins and speed, as they are both sensitive to threshold variations.

Lastly, the area of a GaAs wafer is considerably smaller (about four times) than a Si wafer. Moreover, as shown in Fig. 5.7.3, the power dissipated per stage in a GaAs circuit is large compared to that in a CMOS circuit. We recall that the thermal coefficient of GaAs is relatively high. For all these reasons, the number of transistors that can be packed on a GaAs chip is much smaller than on a Si chip and for now, only medium to large scale integration is practical. The implication on system design is that the system speed performance will be determined by the pulse propagation characteristics of the inter-chip interconnects as well as by the switching speed of the circuits.

With these characteristics in mind, we now present the design of the major subsystems that make up a lightwave digital communications system, as illustrations of the applications of GaAs digital circuits.

7.2 Lightwave Communications System

A simplified lightwave communications system is shown schematically in Fig. 7.2.1. N digital signal streams of M b/s, each of which could be a time-multiplexed signal of several digital signal streams of slower bit rates, are combined into a signal stream of NM b/s by a multiplexer. The electronic signal is converted to an optical signal (E/O conversion) by a semiconductor laser, which launches a lightwave onto the optical fiber. After some distance, the weakened optical signal is converted to an electronic signal (O/E), amplified, reshaped, and converted back to an optical signal (E/O) at a repeater. P such NM b/s signals are carried in a bundle of P fibers. Depending on the system application, each of the P NM b/s signals may be switched into a different fiber (space switching), or each of the N M-b/s signals in a given time slot on a given fiber may be switched into a different time slot on a different fiber (time and space switching). In any case, some kind of cross-point switch is required. The P NM-b/s signals at the output of the cross-point switch are converted into an optical form again. At the receiver, after an O/E conversion, each

NM-b/s signal is de-multiplexed into N M-b/s signals.

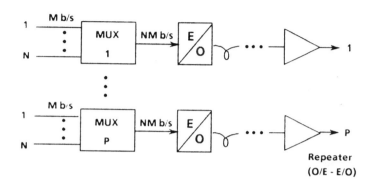

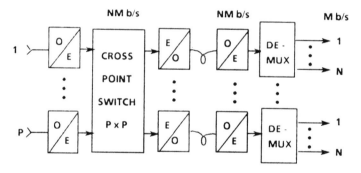

Fig. 7.2.1 A simplified digital lightwave communications system.

The bandwidth of an optical fiber is in the range of tens of Tera Hertz so that potentially a fiber can carry a digital signal of tens Tera-b/s. However, electronic circuits do not exist today that can operate at that kind of speed, so that the bottleneck in a lightwave communications system is the electronics/optics interface. Secondly, until lightwave amplifiers are practical, the regeneration of the optical signal in a long distance system must go through the 2-stage conversion of optics-to-electronics and electronics-to-optics. In both instances, it is clear that high speed electronic circuits are needed.

In the following, we shall describe the design of some of the sub-systems of a lightwave communications system. These are the multiplexer/demultiplexer, cross-point switches, the repeater (detection circuit, clock extraction circuit, decision circuit and laser driver), and a static random access memory. But first we must present an important system requirement that bears directly on the design of the digital circuits that make up the subsystems. The requirement is pulse-width preservation.

7.3 Pulse Width Preservation

In a digital communications system, synchronization of the signals and synchronization of the circuits that process them are of paramount importance. Since it is impractical to distribute a system clock signal throughout the system, timing must be done locally at each system component and the system clock must be derived from the data signal. The data signal is often encoded so that the system clock information can be extracted from it. For example, in a non-return-to-zero (NRZ) scheme [1], the data pulse occupies the entire time period of the system clock, so that the leading edge and the falling edge of a data pulse mark the beginning and end of a system clock period, respectively. These edges are detected in a clock extraction circuit and their separation in time determines the period of a sine wave that reproduces the system clock. If the sine wave generated locally at each repeater is to have the same period as the system clock, the pulse width of the data pulses must be preserved throughout the system. This means that the circuits along a data path must have the property that the output pulse width be the same as the input pulse width.

In a different but similar setting, in many a digital system, the time interval over which the logical value of a data pulse is "valid," namely, the pulse width, should be the same for all data pulses, for otherwise, delay must be inserted to synchronize the occurrence of logical events. The amount of compensating delay will depend on the circuits that process the data pulses and will be different on different data paths. Clearly, this makes the design of the system extremely difficult. It is therefore prudent to require at the outset that throughout the system, pulse width be preserved at all logic gates.

In the following, we will describe the principle of pulse preservation.

Consider two inverters in cascade (Fig. 7.3.1). Let the input waveform $V_A(t)$ be approximated as a trapezoid as shown. We will assume that the output of inverter 1, $V_B(t)$, and of inverter 2, $V_C(t)$, are also trapezoidal with the rising and falling slopes given in the figure. We further assume that the logical 1 level is 1 V and logical 0 is 0 V. Let V_T be the circuit threshold voltage such that when the input to an inverter exceeds V_T on the rising edge, the output begins to fall, and when the input goes below it, the output begins to rise.

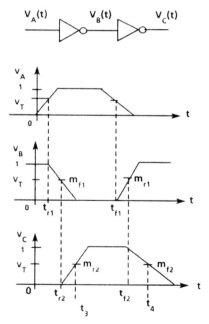

Fig. 7.3.1 Analysis of pulse width preservation.

With reference to the figure, the pulse width of the input is

$$PW_{in} = t_{f1} - t_{r1} \qquad (7.3.1)$$

Pulse $V_C(t)$ begins at t_{r2} and starts to fall at t_{f2}, and these time points are given by

$$t_{r2} = t_{r1} + \frac{1 - V_T}{m_{f1}} \tag{7.3.2}$$

$$t_{f2} = t_{f1} + \frac{V_T}{m_{r1}} \tag{7.3.3}$$

The points at which the pulse width of $V_C(t)$ are defined are given by

$$t_3 = t_{r2} + \frac{V_T}{m_{r2}} \tag{7.3.4}$$

$$t_4 = t_{f2} + \frac{1 - V_T}{m_{f2}} \tag{7.3.5}$$

The pulse width of the output is then

$$PW_{out} = t_4 - t_3 \tag{7.3.6}$$

$$= PW_{in} + V_T(\frac{1}{m_{r1}} - \frac{1}{m_{r2}}) + (1 - V_T)(\frac{1}{m_{f2}} - \frac{1}{m_{f1}})$$

The pulse width can expand or shrink, depending on the values of the rising and falling slopes. For the case in which the rising slopes are the same, $m_{r1} = m_{r2}$, and the falling slopes are the same, $m_{f1} = m_{f2}$, the pulse width at the output equals that of the input. Furthermore, it is independent of the circuit threshold voltage V_T.

In addition, if we define $t_3 - t_{r1}$, as the rising delay of the circuit and $f_4 - t_{f1}$ as the falling delay, then they are equal (necessarily) under the same conditions on the rising and falling slopes of the response of the inverters.

The significance of pulse width preservation is that the duration in which a signal is "valid" is unchanged from point to point in a logic chain so that in a tightly timed circuit, the time interval between logic events need not be stretched at any point along the chain. Thus, the circuit can be operated at a speed determined by the input data rate and not by the circuit, except for a predictable, constant delay through the circuit.

7.4 Time-Multiplexer/Demultiplexer

A time-multiplexer is a circuit that combines N data bit-streams of M b/s into a single bit-stream of NM b/s, with successive time slots allotted to the data bits of successive streams. A schematic diagram of a 4:1 multiplexer and an implementation using NOR gates are shown in Fig. 7.4.1.

The four data streams are sampled successively by the control signals X and Y. An extra NOR gate is inserted along the combining path Each NOR gate of the same pair must have the same pull-up and pull-down delays (more precisely, the same rising and falling slopes) in order that the pulse width of the signal is preserved as the signal propagates from the input to the output. Moreover, by making the rising and falling slopes of the NOR gates of a pair equal, the pulse width is insensitive to the changes of the circuit threshold from pair to pair, as explained in the last section.

The control signals that sample the N bit streams successively are generated by a counter clocked at a frequency of NM b/s. As shown in the figure, the counter consists of two D-flip-flops, which can be of the 6-NOR type if ED or buffered ED logic is used, or of the series-gating type if source-coupled logic is used.

As is well-known in digital circuit design, if two supposedly coincident pulses are combined at a NOR gate, with one slightly delayed (advanced) with respect to the reference pulse, a positive pulse is generated at the output at the time of the trailing (leading) edge of the reference pulse. This unwanted positive pulse ("glitch") can be removed by a conventional "de-glitch" circuit shown at the output of the multiplexer.

Lastly, Fig. 7.4.2 shows a simple design of a de-multiplexer. The input bit stream of NM b/s is applied to N (N=4 in the figure) D-flip-flops, each being clocked in time sequence. At the 4th timing pulse, the output of each flip-flop is latched onto a register, and the 4 bits are taken in parallel in space. Note that the Nth timing pulse is applied to N latches and should be buffered.

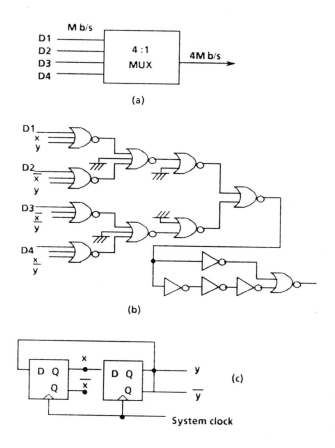

Fig. 7.4.1 A 4:1 time-multiplexer.

7.5 Cross-Point Switch

In Sec. 4.4, we presented a simple design of a cross-point switch based on transmission gates. We shall now consider other designs which have improved performance.

Multiplexer-Based Design

We recall that in a cross-point switch, each output, independent of the other outputs, can receive data from one and only one input. This suggests that in the case of a 4x4 switch, we use four independent 4:1

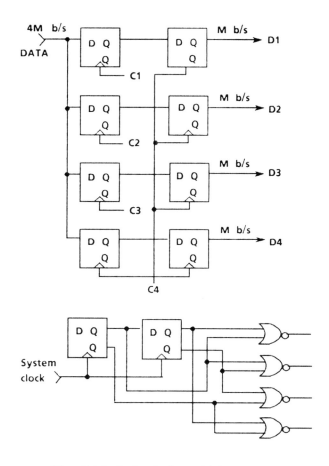

Fig. 7.4.2 A simple de-multiplexer.

multiplexers as shown in Fig. 7.5.1. At any one time, only one of the four inputs is selected at a multiplexer.

The multiplexer, shown in Fig. 7.5.2, is simliar to the time-multiplexer of the last section. To preserve pulse width, the 2-input NOR gates by pairs must have matching rising and falling slopes. Note that each input is applied to all of the multiplexers and must be buffered. For this reason, the speed of a cross-point switch is often determined not so much by the speed of the multiplexers as by the delay of the buffers. In a large switch, each input would be distributed to the multiplexers in a tree-like structure, with a buffer at each node of the tree.

The control signals c_{ij} will be generated by 4 decoders shown schematically in Fig. 7.5.3. Only one out of four control signals is active at any one time at a given decoder, as determined by the Boolean value of the switching signals s_{1i} and s_{2j}.

In a 4x4 switch, there are 16 control signals, c_{ij}, i,j=1,...,4, which can be regarded as a 16-bit word. The Boolean value of the word at a given time specifies the switching pattern of the switch. Since the output can come from any of the 4 inputs, and there are 4 independent output terminals, there are 4^4=256 different switching patterns. These patterns, instead of being generated by the decoders, can be stored in a 256-word 16-bit/word memory, accessed by an 8-bit address. Depending on the application, the switching pattern may change infrequently, or it may change as fast as the bit rate.

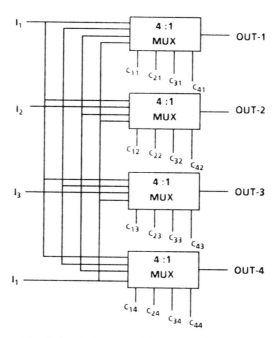

Fig. 7.5.1 A 4x4 multiplexer-based cross-point switch.

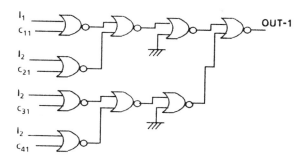

Fig. 7.5.2 A NOR gate implementation of a multiplexer.

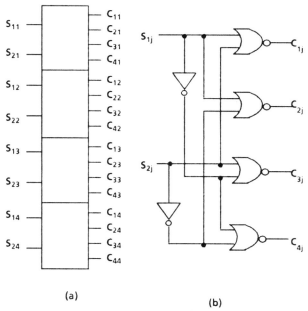

(a)

(b)

Fig. 7.5.3 The decoder (a) and an NOR implementation (b).

Transmisson-Gate Based Design

As shown in Sec. 4.4, a cross-point switch can also be realized as a matrix of switch cells. Each cell is made of transmission gates. In the earlier design, the output voltage has a maximum value equal to one threshold voltage drop from the gate voltage. Moreover, when the output is low, the control voltage may leak to the output through the gate-to-source diode of the series transistor, causing the output to attain a value which could be interpreted as a logical 1.

These problems can be avoided in a better design shown in Fig. 7.5.4. The transistors are all depletion type with a threshold voltage of -1 V. When the control voltage $C=0V$ and $\overline{C}=-1V$, transistors T_2 and T_5 are both cut off. Since the gate and source of T_1 are connected together, $V_1=V_{in}$. As a result, the gates of T_3 and T_4 are effectively connected to the input and $V_2=V_{in}$ and by extension $V_{out}=V_{in}$ in the steady state. When $C=-1V$ and $\overline{C}=0V$, T_2 and T_5 are both conducting, V_1 is close to $-1V$ and V_2 is close to $0V$. T_3 and T_4 are cut off and the switch cell is disconnected from the output bus, as required.

Note that the control signal is isolated from the data signal path. The leakage of the control signal into the output is through interelectrode capacitive coupling and is noticeable only when the control signal changes from one value to another.

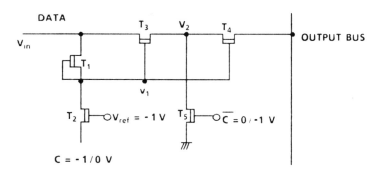

Fig. 7.5.4 A transmission-gate based cross-point switch cell.

The switch cell of Fig. 7.5.4 has one deficiency in that when $C=-1V$, T_2 conducts and there is a conducting path from the input source to the control signal source C. The circuit that generates C must be able to sink this current without changing the voltage at C. In other words, C should come from an ideal voltage source. To circumvent this require-ment, the circuit of Fig. 7.5.5 is proposed [2]. In it, transistors T_1 and T_2

constitute a voltage divider, which is a constant load on the voltage source of C. Assume the threshold voltage of the transistors is -1 V. When $C=0$ and $\bar{C}=-1.5V$, T_3 and T_5 are cut off. V_1 is at 0 V. If the input signal swings between 0 and 1 V, T_4 conducts and V_3 and hence V_{out} follows the input. When $C=-3V$, $V_2=-1.5V$, T_3 conducts and V_3 is at -1.5 V. Since $V_1=-3V$, T_4 is cut off, as are T_6 and T_7. At the same time, V_5 is at ground and the cell is disconnected from the output bus.

In both of the transmission-gate based designs, the control signal C and its complement $\bar{C}$ are needed at each cell. This suggests the use of a source-coupled differential pair to generate them from an input c_{ij}. To avoid the necessity of carry C and $\bar{C}$ on long conductors, each differential pair should be placed next to a cell. The reference voltage V_{ref} should be generated locally as well, if possible, as shown in Fig. 7.5.5(b).

7.6 Time- and Time-Space-Switches

A time-switch is a circuit that takes an input bit stream consisting of frames of N-bit data in N time slots and produces an output stream of the same form except that in each output time slot is placed an input data bit from any one of the N input time slots, in accordance with a switching control signal applied to the switch at a rate equal to the frame rate [3]. Thus, a data item in a time slot is switched into another one or more, or none, of the time slots in the output stream. Fig. 7.6.1 gives the functional description of the switch. Such a switch is used in an ISDN to provide the same switching function as a cross point switch except the latter switches in space whereas the time-switch re-directs the data in time.

In this example, each frame consists of 8 bits in 8 time slots. The switch places the bit in input time slot T4 into output time slot 1, T6 into output time slot 2, T1 into time slots 3 and 4, etc. A simple implementation of the switch is shown in Fig. 7.6.2.

As shown in the figure, the input data stream is latched and applied to an 8-bit shift register. When all 8 bits of a frame are in place, the 8-bit data are transferred to the 8-bit latch. Any one of the 8 data bits can be selected by the 3-bit select signal applied to the 8:1 selector and it is latched at the output. By changing the 3-bit select signal at the bit rate, the 8 data bits are placed at the output time slots, one at a time, as

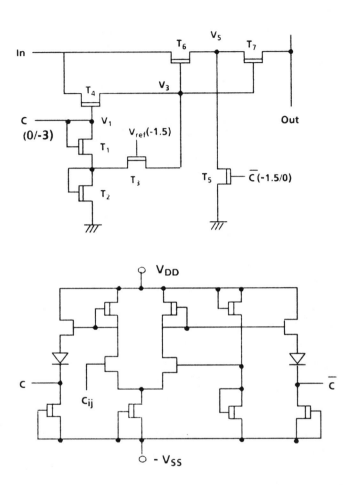

Fig. 7.5.5 (a) A second design of the switch cell. (b) Generation of the
control signal and its complement of cell (i,j).

specified in the switching pattern, to form an output frame. The 8 3-bit
select signals (words) are stored in a RAM. The READ address is generated by a modulo-8 counter at the bit rate, which also provides the load
pulse at the frame rate to latch the 8 input bits from the shift register.

The time-switch can be replicated in space to become a time-
space-switch. Fig. 7.6.3 shows an example with 8 input channels and 8
output channels. Each input stream consists of frames of 8 data bits in 8
time slots. Any of the 64 data bits can be switched to any of the 64

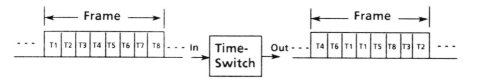

Fig. 7.6.1 Functional description of a time-switch.

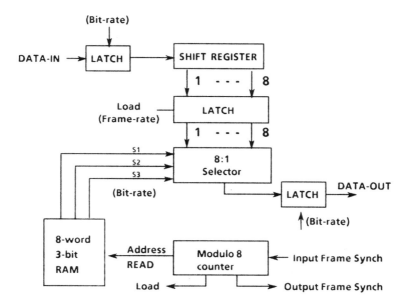

Fig. 7.6.2 Block diagram of a time-switch.

output time-space slots. An implementation based on the time-switch of Fig. 7.6.2 is shown in Fig. 7.6.4. The shift register and the 8-bit latch are replicated 8 times. The 3-bit select signal selects a "column" of the bank of 8 8-bit latches. The 8 output bits from the latches are applied to the 8:1 selector. One of the 8 bits is then selected by the 3-bit select signal S4, S5, S6 and latched at the output. Eight 6-bit select words are stored in a RAM which is addressed by a 3-bit address generated by a modulo-8 counter, as before.

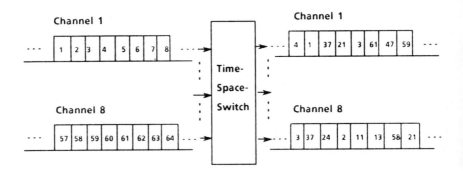

Fig. 7.6.3 Functional description of a time-space-switch.

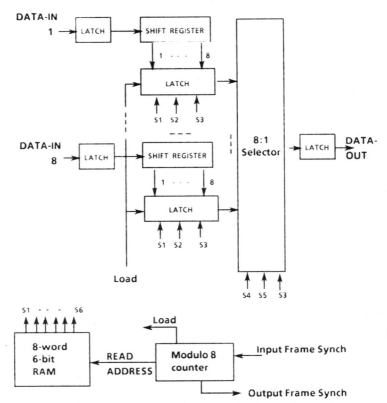

Fig. 7.6.4 Block diagram of a time-space-switch.

7.7 Static Random Access Memory

Depending on the system application, the usual design criteria of a static random access memory (SRAM) are density (the number of cells per chip), power consumption, and read and write access times. To maximize the density, the circuits that make up the cells, address decoders, and sense amplifiers should be as simple as possible. To minimize power dissipation, those circuits that are idle at a given time should not draw any current. To minimize the access times, the pull-up and pull-down delays of the circuits should be small. Lastly, there should preferrably be only one supply voltage, for otherwise, additional power busses will be needed and they take up chip area. All these considerations point to the ED logic as the circuit of choice for a SRAM [7.4, 7.5].

Fig. 7.7.1 shows a functional block diagram of one cell of a SRAM. Its structure is conventional. The data bit is stored in the ED-latch. Cells along a row are accessed through the word line (WL) transistors T_1 and T_2 activated by the X-address circuit. The value of the stored bit and its complement appear along the bit line (BL) and ($\overline{BL}$), respectively, which are connected to all the cells in the same column. BL and $\overline{BL}$ are accessed through the transistors T_3 and T_4 activated by BLS (bit line select) of the Y-address circuit. To write a bit into the cell, the write-enable signal WE is high and the data bit and its complement appear on BL and $\overline{BL}$. Transistors T_1 through T_4 are turned on by WL and BLS and the data bit is placed in the cell. To read, the pass transistors are activated in the same way and WE goes low to force the output of the write circuit to be "floating", and at the same time, the read circuit is enabled by $\overline{WE}$ to generate an output that represents the value of the stored bit.

Address Circuit

Fig. 7.7.2 shows the X-address circuit. It consists of an ED NOR gate as a decoder followed by an EDSFD buffer. The value of the output WL must be within the allowable range specified by the inequality of Sec. 4.3. WL must drive a number of transmission gates equal to twice the number of cells in a row and the load is principally capacitive. When a row is not selected, WL is low and the last buffer is conducting. Since it is the largest in area, it carries the most current, and the idle power is

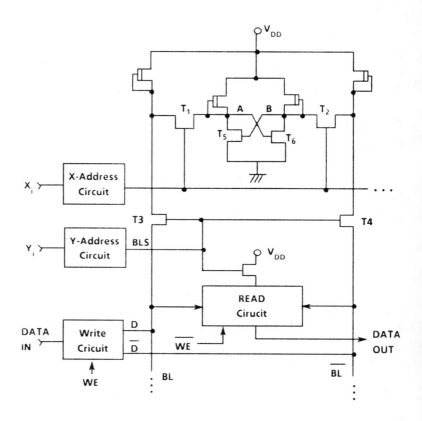

Fig. 7.7.1 Functional block diagram of a SRAM.

mainly determined by the idle power of the last buffer. The Y-address circuit is similar except the output drives only three transmisison gates, T_3, T_4 and one that connects the supply voltage to the read circuit.

READ Circuit

The READ circuit is simply a sense amplifier. There are many possible designs. Fig. 7.7.3 shows one based on ED logic and it consists of a pair of push-pull amplifiers driven by BL and $\overline{BL}$. The output of the pair drives a super buffer. To conserve power when the column is not selected, its sense amplifier is disconnected from the supply through the pass transistor T_7 which is de-activated by BLS, the output of the Y-

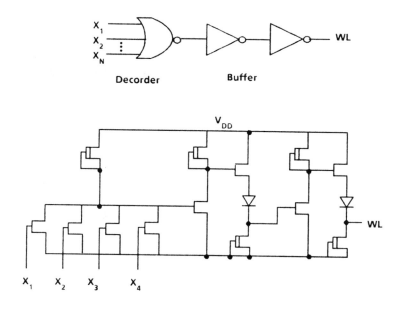

Fig. 7.7.2 X-address circuit.

address circuit.

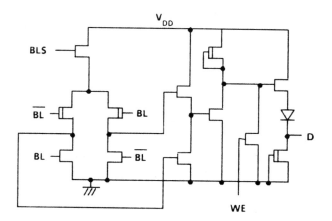

Fig. 7.7.3 The READ circuit.

WRITE Circuit

Fig. 7.7.4 shows the WRITE circuit. The data and its complement each drives a super buffer. When write-enable WE is low, the transistors of the super buffers are open and the circuit is disconnected from the bit lines BL and $\overline{BL}$.

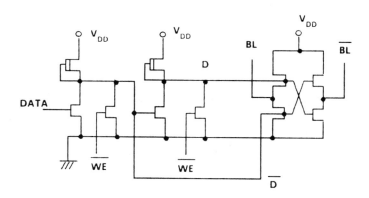

Fig. 7.7.4 The WRITE circuit.

Improvements

The basic SRAM cell design can be improved. With reference to Fig. 7.7.1, suppose the cell is not selected so that WL is low. Assume node A is low so that T_5 is conducting. If T_1 is not perfectly cut off, as is the case when its gate to source voltage is near the threshold voltage , there will be leakage current through the bit line pull-up transistor T_8, and the transistors T_1 and T_5. If the leakage current is substantial, the voltage at A could be high enough to cause a logic error. In one improved design [7.6], the leakage current is reduced by connecting the common source of the latch to a parallel connection of an enhancement transistor and diode, as shown in Fig. 7.7.5. When the cell is selected, node S is at ground, as required. When the cell is not selected, the voltage at S, and at L, is raised. The gate to source voltage of T_1 is then well below the threshold voltage of the trasistor and the leakage current will be very small.

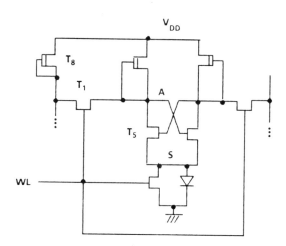

Fig. 7.7.5 A means to reduce leakage current through the access transistor.

In another design [7.7], a bootstrap circuit is used in place of the pull-up transistor on a bit line, as shown in Fig. 7.7.6. When a bit, say 1, stored in a cell is transferred through the access transistor T_1 onto the bit line, voltage V_1 rises, forcing the current in T_8 to decrease. However, the presence of the bootstrap capacitor C_{BS} causes voltage V_2 to rise in step with the rise of V_1 so that the gate to source voltage of T_8 is kept unchanged and the charging current on the bit line capacitor C_{BL} is not decreased.

7.8 O/E-E/O Repeater

With reference to the lightwave communications system of Fig. 7.2.1, the optical signal on a fiber, after travelling some distance, becomes smaller in amplitude because of energy loss in the fiber, and it is also distorted owing to dispersion. The signal must be re-generated periodically at repeaters placed along the length of the fiber system. At a repeater, the light signal is first converted to an electronic signal at the data bit rate, and then re-converted to a light signal and re-launched onto the fiber. Fig. 7.8.1 shows the block diagram of a repeater. The light signal is detected by a photodiode of the junction type (PIN) or the avalanche type [8]. The diode current is applied to a current-to-voltage

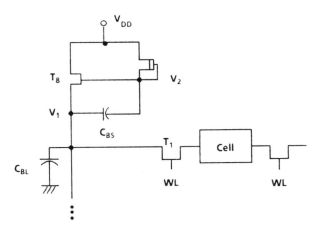

Fig. 7.7.6 Bootstrap circuit to improve READ access time.

converter (I/V). Its output is applied to a decision circuit, whose function is to determine if the input pulse is a logical one or zero. The system clock is extracted at the extraction circuit which provides timing signals for the decision circuit. The output of the decision circuit is amplified and is used to modulate a semiconductor laser, which produces an light pulse at the output.

The individual circuits will now be described in fuller detail.

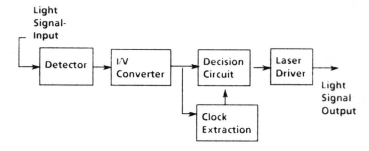

Fig. 7.8.1 Block diagram of an O/E-E/O repeater

Detector Circuit

Fig. 7.8.2 shows the circuit of a detector. The photodiode is biased so that electron-hole pairs generated by the incident light energy will produce electron and hole currents that combine additively as output current. The current is converted to a voltage at the current-voltage converter, which is basically a high gain amplifier with a feedback resistor connected to the input. The op-amp should have high input impedance, low output impedance, and high gain. One possible design is shown in Fig. 7.8.3. The static or dc current-voltage transfer curve is shown in the figure and is seen to be approximately linear over the range of 0 to 1 mA.

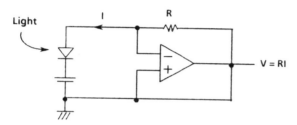

Fig. 7.8.2 A detector circuit.

Clock Extraction Circuit

Fig. 7.8.4 shows the elements of a clock extraction circuit. The input pulses are differentiated and rectified, and the resultant pulses are applied to a resonant circuit which produces a damped oscillation at the system clock frequency. Provided that there are no long periods of all 1's and 0's in the input pulse stream, the oscillation can be sustained. It is then amplified and clipped to produce a square wave with the same period as the system clock. The waveforms of the processed signal at the various stages are shown in the figure.

Decision Circuit

The decision circuit is essentially a one-bit quantizer. When the input exeeds some threshold value, the output is high; otherwise it is low. This suggests that we use a comparator followed by a latch. In one design [9], shown in Fig. 7.8.5, the comparator is a source-coupled

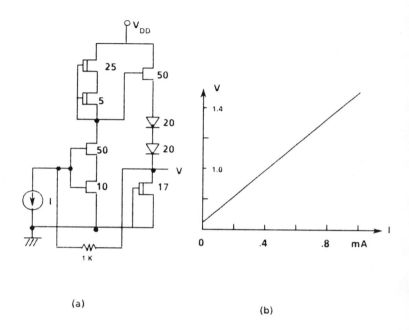

(a)

(b)

Fig. 7.8.3 Current-voltage converter and its transfer characteristic.

differential pair and the latch is a source-coupled flip-flop which shares the same load transistors as the differential pair. In this way, the two circuits are tightly coupled. Latching occurs on a positive clock pulse.

In a second design, shown in Fig. 7.8.6 [10], the comparator is a high gain cascode differential pair discussed in Sec. 6.6. The output of the comparator D and $\overline{D}$ are used to control the effective resistance of the pass transistors T_1 and T_2, respectively. If D is high, the "resistance" of T_1 is low and node A is high. At the same time, the "resistance" of T_2 is high and node B is low. At the output of the source follower, Q is high and $\overline{Q}$ is low, provided that *clock*, which is applied to both legs, is low. In this case, latching occurs when the clock goes from high to low.

Laser Driver

A semiconductor laser emits light by virtue of stimulated emission of photons induced by an external bias current applied across the active layer of the device [11]. By modulating this bias current in response to a

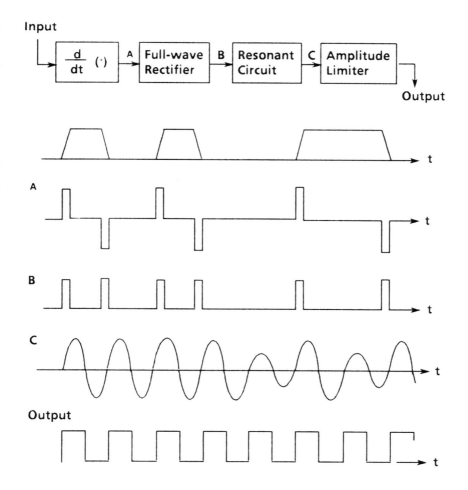

Fig. 7.8.4 A clock extraction circuit and the associated waveforms.

signal, the intensity of the lightwave is modulated to carry the signal information. In an InGaAsP laser of wavelength of 1.3 μM, a bias current of about 70 mA and a modulating current of 40 mA are required to achieve 100 % modulation of the light intensity [12]. A laser driver must therefore supply the bias current and provides a means to add the signal current to the bias current.

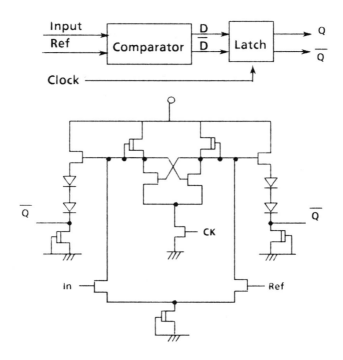

Fig. 7.8.5 A decision circuit [9].

A simple design of the laser driver is shown in Fig. 7.8.7. Transistors T_1 and T_2 constitute a source follower with the laser as a load. The bias current is supplied by a parallel connection of several current sources $Q_1, ..., Q_n$.

A second design is shown in Fig. 7.8.8. The laser is a load of one leg of the output differential pair. Note that because the laser requires a relatively large amount of current to achieve full modulation, the transistors of the differential pair are very large. Since it is not practical to fabricate a transistor with a width of larger than 200 μM, each transistor is usually realized by a parallel combination of several small transistors.

In this circuit, the bias current is supplied by a current mirror, shown in Fig. 7.8.9. The bias current is equal to the reference current multiplied by the ratio of the transistor widths W_B / W_R.

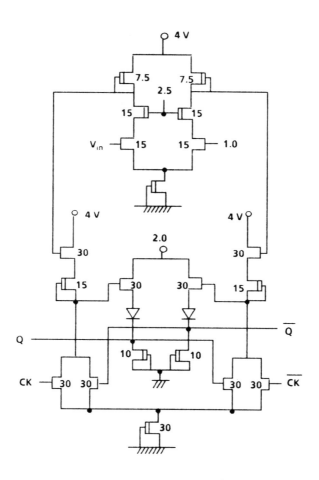

Fig. 7.8.6 A one-bit quantizer [10].

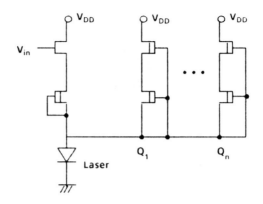

Fig. 7.8.7 A simple design of a laser driver.

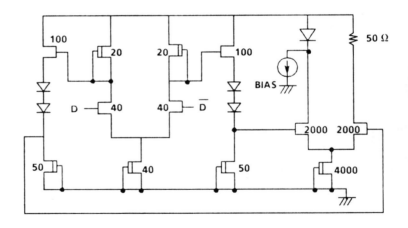

Fig. 7.8.8 A second design of the laser driver.

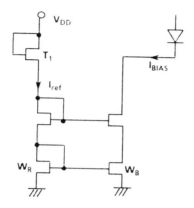

Fig. 7.8.9 A current mirror to provide the bias current to a
semiconductor laser.

7.9 Remarks

In this chapter, we have presented the design of the major subsystems of a lightwave communications system. We discussed the importance of pulse width preservation and showed how it can be theoretically achieved. Unfortunately, pulse width preservation remains an objective that is difficult to meet in practice, since there is no simple way to relate the actual pulse width of the pulse response of a circuit to the design parameters such as the transistor widths. Much trial and error is needed to obtain a design whose rising and falling slopes meet the specified value. This problem merits further study.

Cross-point switches are important not only in telecommunications systems but also in multiprocessor computer systems where each processor must be able to communicate with any other and also with any one of several memory modules. As to whether the multiplexer-based or the transmission-gate based design is preferred, much depends on the application. If the number of ports N is small, then the transmission-gate based design is adequate. However, in the latter design, the input signal line must drive a cell whose output is connected to a bus loaded by $N-1$ other pass transistors. Moreover, if the input is broadcast to several cells, the load is increased by several folds. So for large N (≥ 16), the multiplexer-based design is a better choice. Additional references are found in [13 - 17].

We have already mentioned the large-size transistors that are needed in a laser driver in order to provide the necessary bias and modulating currents. Recent developments in MESFET and HFET technology show that both the needed speed and current capacity are now available at as high as 10Gb/s rates [18 - 19].

There are other GaAs subsystems that are used in high speed communications and computer systems. Among them are (1) A/D and D/A converters operating at Giga-samples per second that are needed in radar and high definition TV systems [20 - 31]; (2) programmable and multi-modulus prescalers used in frequency synthesizers [32 - 42]; (3) ALUs and parallel multipliers for digital signal processing applications [43 - 46]; and (4) gate arrays for general applications [47 - 49]. Because of space limitations, these circuits will not be presented. In all cases, the basic logic building blocks are the inverter and the NOR gate. For low-power applications, the ED logic is used. For circuits where noise margins are important, the EDSF or its variation EDSFD and the source-coupled differential pair are preferred.

As a last remark, we should mention that heterojunction field-effect transistors (HFET) with submicron gate lengths and heterojunction bipolar transistors (HBT) are being developed that will soon find themselves used in digital circuits that have operating speeds in the tens of GHz range. The HFET has current-voltage characteristics similar to those of the MESFET, and the circuit design techniques developed in this book for MESFET will be directly applicable to HFET circuits. As to HBT circuits, the design techniques for Si bipolar transistors seem to carry over with little modifications, as exemplified in the design of a 34.8 GHz frequency divider [50], time multiplexer switch [51], and DC to 10 GHz direct-coupled amplifier [52].

References

[1]   F.F.E. Owen, **PCM and Digital Transmission Systems**. New York: McGraw-Hill Book Company, 1982. Chapter 8.

[2] Defu Luan and O. Wing, "Design of a GaAs 4x4 cross-point switch," Proceedings of International Conference on Circuits and Systems, Nanjing, China, 1989, pp. 53-56.

[3] Y. Shimazu and T. Takada, "High speed time switch using GaAs LSI technology," IEEE Journal on Selected Areas in Communications, Vol. SAC-4, No. 1, January 1986, pp. 32-38.

[4] T. Mizoguchi, N. Toyoda, K. Kamazawa, Y. Ikawa, T. Terada, M. Mochizuki and A. Hojo, "A GaAs 4K bit static RAM with normally on and off combination circuit," IEEE GaAs Integrated Circuit Symposium Technical Digest, 1984, pp. 117-120.

[5] S. Takano, H. Makino, N. Tanino, M. Noda, K. Nishtani and S. Kayano, "A GaAs 16K SRAM with a single 1-V supply," IEEE Journal of Solid State Circuits, Vol. SC-22, No. 5, October 1987, pp. 699-703.

[6] H. Makino, S. Matsue, M. Noda, N. Tanino, S. Takano, K. Nishitani and S. Kayano, "A 7ns/850mW GaAs 4Kb SRAM fully operative at $75^{o}C$, " IEEE GaAs Integrated Circuit Symposium Technical Digest, 1988, pp. 71-74.

[7] T. Hayashi, H. Tanaka, H. Yamashita, N. Masuda, T. Doi, J. Shigata, N. Kotera, A. Masaki and N. Hashimoto, "Small access time scattering GaAs SRAM technology using bootstrap circuits," IEEE GaAs Integrated Circuit Symposium Technical Digest, 1985, pp. 199-202.

[8] S. Sze, **Physics of Semiconductor Devices**. New York: Wiley Interscience, 1981, p. 749.

[9] D. Heigmant and N. Binet, "A high performance 1.8 GHz strobed comparator for A/D converter," IEEE GaAs Integrated Circuit Symposium Technical Digest, 1983, pp.66-69.

[10] T. Nguyen, F. Thomas, S. Ruggeri, M. Le Paih, J.M. Uro, F. Debrie, I. Dean and M. Gloanec, "A 4-bit full Nyquist 1 G Sample/s monolithic GaAs ADC with on-chip S/H and error correction," IEEE GaAs Integrated Circuit Symposium Technical Digest, 1988, pp. 195-198.

[11] E. S. Yang, **Microelectronic Devices**. New York: McGraw-Hill Book Company, 1988, pp. 404-405.

[12] J. E. Bowers, "High speed semiconductor laser design and performance," Solid State Electronics, Vol. 30, No. 1, 1987, pp. 1-11.

Cross-point Switches

[13] C. J. Andersen, "A GaAs MESFET 16x16 crosspoint switch at 1700 Mb/s," IEEE GaAs Integrated Circuit Symposium Technical Digest, 1988, pp. 91-94.

[14] H. M. Park and H. C. Ki, "A GaAs MESFET 16x8 crosspoint switch," IEEE International Solid State Circuits Conference Digest of Technical Papers, 1989, pp. 146-147.

[15] T. Iakada, Y. Shimazu, K. Yamasaki, M. Togashi, K. Hoshikawa and M. Idda, "A 2 Gb/s throughput GaAs digital time switch LSI using LSCFL," IEEE Microwave and Millimeter Wave Monolithic Circuit Symposium Proceedings, 1985, pp. 22-26.

[16] S. Hayano, K. Nagashima, S. Asai, T. Maeda and T. Furutsuka, "A GaAs 8x8 matrix switch LSI for high-speed digital communications," IEEE GaAs Integrated Circuit Symposium Technical Digest, 1987, 00. 145-148.

[17] G.W. Dick, R. F. Huisman, Y. K. Jhee, R. A. Nordin, W. A. Payne and K. W. Wyatt, "2.4 Gb/s GaAs 8x8 time multiplexed switch integrated circuit," IEEE GaAs Integrated Circuit Symposium Technical Digest, 1989, pp. 101-104.

Laser Drivers

[18] N. Kotera, K. Yamashita, Y. Hatta, T. Kinoshita, M. Miyazaki and M. Maeda, " Laser Driver and receiver amplifiers for 2.4 Gb/s optical transmission using WSi-GaAs MESFET's," IEEE GaAs Integrated Circuit Symposium Technical Digest, 1987, pp. 103-106.

[19] Y. Suzuki, H. Hida, T. Suzaki, S. Fujita, Y. Ogawa, A. Okamoto, T. Toda and T. Nozaki, "A 10-Gb/s laser driver IC with i-AlGaAs/n-GaAs doped-channel hetero-MISFETs (DMTs)," IEEE

GaAs Integrated Circuit Symposium Technical Digest, 1989, pp. 129-132.

A/D and D/A

[20] K. de Graaf and K. Fawcett, "GaAs technology for analog-to-digital conversion," IEEE GaAs Integrated Circuit Symposium Technical Digest, 1986, pp. 205-208.

[21] G.S. LaRue, "A GHz GaAs digital to analog converter," IEEE GaAs Integrated Circuit Symposium Technical Digest, 1983, pp. 70-3.

[22] K-C Hsieh, T. A. Knotts and G. L. Balwin, " A GaAs 12-bit digital to analog converter," IEEE GaAs Integrated Circuit Symposium Technical Digest, 1985, pp. 187-190.

[23] T. Ducourant, J. Baelde, M. Binet and C. Rocher, "1-GHz, 16mW, 2-bit analog to digital GaAs converter, " IEEE Journal of Solid State Circuits, Vol. SC-21, No. 3, June 1986, pp. 453-456.

[24] T. Ducourant, M. Binet, J. Baelde, C. Rocher and J. Gibereau, "3 GHz, 150mW, 4-bit GaAs analog to digital converter," IEEE GaAs Integrated Circuit Symposium Technical Digest, 1986, pp. 209-212.

[25] T. Ducourant, D. Meignant and P. Bertsch, " A 5-bit, 2.2 Gs/s monolithic A/D converter with GigaHertz bandwidth, and 6-bit A/D converter systems," IEEE GaAs Integrated Circuit Symposium Technical Digest, 1989, pp. 337-340.

[26] F. Weiss, "A 1Gs/s 8-bit GaAs DAC with on-chip current sources," IEEE GaAs Integrated Circuit Symposium Technical Digest, 1986, pp. 217-220.

[27] J. Naber, H. Singh and R. Sadler, "A low-power, high-speed 4-bit GaAs ADC and 5-bit DAC," IEEE GaAs Integrated Circuit Symposium Technical Digest, 1989, pp. 333-336.

[28] J. Kleks, C. Robertson, M. Englekirk, K. Fawcett, C. Saunders, M. Listvan, K. Tan and H. Chung, "A 4-bit single chip analog to digital converter with 1.0 GHz analog input bandwidth," IEEE GaAs Integrated Circuit Symposium Technical Digest, 1987, pp. 79-82.

[29] K. Wang, P. Asbeck, M. Chang, G. Sullivan and D. Miller, "A 4-bit quantizer implemented with AlGaAs/GaAs heterojunction bipolar transistors," IEEE GaAs Integrated Circuit Symposium Technical Digest, 1987, pp. 83-86.

[30] B. Wong and K. Fawcett, "A precision dual bridge GaAs sample and hold," IEEE GaAs Integrated Circuit Symposium Technical Digest, 1987, pp. 87-90.

[31] A. Oki, M. Kim, J Camou, C. Robertson, G. Gorman, K. Weber, L. Hobrock, S. Southwell and B. Oyama, "High performance GaAs/AlGaAs heterojunction bipolar transistor 4-bit and 2-bit A/D converters and 8-bit D/A converter," IEEE GaAs Integrated Circuit Symposium Technical Digest, 1987, pp. 137-140.

Prescalers and Dividers

[32] S. Shimizu, Y. Kamatani, N. Toyoda, K. Kanazawa, M. Mochizuki, T. Terada and A. Hojo, "A 1 GHz 50 mW GaAs Dual modulus divider IC," IEEE Journal of Solid State Circuits, Vol. SC-19, No. 5, October 1984, pp. 710-715.

[33] M. Rocchi, B. Chantepie, B. Gabillard, G. Haldermann, J. Debost, J. Andriex and F. Robert, "A 1.2 GHz frequency synthesizer using a GaAs 20/21/22/23/24 modulus divider," IEEE International Solid State Circuits Conference Digest of Technical Papers, 1985, pp. 26-27.

[34] Y. Kamatani, S. Shimizu, N. Uchitomi, K. Kawakyo, M. Mochizuki, and A. Hojo, "Divide by 128/129 5 mW 400 MHz band GaAs prescaler IC," IEEE GaAs Integrated Circuit Symposium Technical Digest, 1985, pp. 179-182.

[35] S. Saito, T. Takada and N. Kato, "A 5 mW 1 GHz GaAs dual-modulus prescaler IC," IEEE Journal of Solid State Circuits, Vol. SC-21, No. 4, August 1986, pp. 538-543.

[36] T. Sugeta, "A 13 GHz GaAs dynamic frequency divider and prescaler IC," IEEE International Solid State Circuit Conference Digest of Technical Papers, 1985, pp. 198-199.

[37] J. Jensen, L. Salmon, D. Deakin and M. Delancy, "26-GHz GaAs room temperature dynamic divider circuit," IEEE GaAs Integrated Circuit Symposium Technical Digest, 1987, pp. 201-204.

[38] H. Singh, R. Sadler, A. Geissberger, W. Tanis and E. Schineller, "High-speed, low-power GaAs programmable counters for synthesizer applications," IEEE GaAs Integrated Circuit Symposium Technical Digest, 1987, pp. 269-272.

[39] M. G. Kane, P. Y. Chan, S. S. Cherensky and D. C. Fowlis, "1.5-GHz programmable divide-by-N GaAs counter," IEEE Journal of Solid State Circuits, Vol. SC-23, No. 2, 1988, pp. 480-484.

[40] J. F. Jensen, U. K. Mishra, A. S. Brown, R. S. Beaubien, M. A. Thompson and L. M. Jelloian, "25 GHz static frequency dividers in AlGaAs/GaAs HEMT technology," IEEE International Solid State Circuits Conference Digest of Technical Papers, 1988, pp. 266-267.

[41] K. Fujita, H. Itoh and R. Yamamoto, "A 15.6 GHz commercially based 1/8 GaAs synamic prescaler," IEEE GaAs Integrated Circuit Symposium Technical Digest, 1989, pp. 113-116.

[42] I. Saito, H.I. Fujishiro, T. Ichioka, K. Tanaka, S. Nishi and Y. Sano, "0.2 μM gate inverted HEMTs for an ultra-high speed DCFL frequency divider," IEEE GaAs Integrated Circuit Symposium Technical Digest, 1989, pp. 117-120.

ALUs and Multipliers

[43] W. R. Leung, Y. K. Lo, Y. T. Oh, W. A. Oswald, E. K. Poon, C. E. Reid, L. E. Ackner and T. C. Poon, "2K gate circuits with 125 ps gate delay using GaAs HFET technology," IEEE GsAs Integrated Circuit Symposium Technical Digest, 1989, pp. 57-60.

[44] C. G. Ekroot, and S. I. Long, "A GaAs 4-bit adder-accumulator for direct digital synthesis," IEEE Journal of Solid State Circuits, Vol. SC-23, No. 2, April 1988, pp. 573-580.

[45] E. Delhaye, C. Rocher, M. Fichelson and L. Lecuru, "A 3.0 ns, 350 mW, 8x8 Booth's multiplier," IEEE GaAs Integrated Circuit Symposium Technical Digest, 1987, pp. 249-252.

[46] H. Singh, R. Sadler, J. Irvine and G. Gorder, "GaAs low-power parallel multipliers for a high-speed digital signal processor," IEEE GaAs Integrated Circuit Symposium Technical Digest, 1987, pp. 253-256.

Gate Arrays

[47] K. Kajii, Y. Watanabe, M. Suzuki, I. Hanyu, M. Kosugi, K. Odani, T. Mimura and M. Abe, " A 40-ps high electron mobility transistor 4.1K gate array," IEEE Journal of Solid State Circuits, Vol. SC-23, No. 2, April 1988, pp. 485-489.

[48] A. Peczalski, G. Lee, W.R. Betten, H. Somal, M. Plagens, J. R. Biard, I. Burrows, B. K. Gilbert, R. L. Thompson, B. A. Naused, S. M. Karwoski, M. L. Samson and S. K. Zahn, "A 6K GaAs gate array with fully functional LSI personalization," IEEE Journal of Solid State Circuits, Vol. SC-23, No. 2, April 1988, pp. 581-590.

[49] G. Lee, S. Canaga, B. Terrell and I. Deyhimy, "A high performance GaAs gate array family," IEEE GaAs Integrated Circuit Symposium Technical Digest, 1989, pp. 33-36.

[50] Y. Yamauchi, O. Nakajima, K. Nagata, H. Ito and T. Ishibashi, "A 34.8 GHz static frequency divider using AlGaAs/GaAs HBTs," IEEE Integrated Circuit Symposium Technical Digest, 1989, pp. 121-124.

[51] J. Parton, P. Topham, D. Taylor, N. Hiams, A. Holden and R Goodfellow, "An HBT programmable hexadecimal counter IC clocked at 2.6GHz," IEEE GaAs Integrated Circuit Symposium Technical Digest, 1989, pp. 61-64.

[52] K. Kobayashi, R. Esfandiari, A. Oki, D. Umemoto, J. Camou and M. Kim, "GaAs heterojunction bipolar transistor MMIC DC to 10 GHz direct-coupled feedback amplifier," IEEE GaAs Integrated Circuit Symposium Technical Digest, 1989, pp. 79-82.

Index